Confronting the Enigma of Time

Confronting the Enigma of Time

John R. Fanchi
Texas Christian University, USA

World Scientific

NEW JERSEY • LONDON • SINGAPORE • BEIJING • SHANGHAI • HONG KONG • TAIPEI • CHENNAI • TOKYO

Published by

World Scientific Publishing Europe Ltd.
57 Shelton Street, Covent Garden, London WC2H 9HE
Head office: 5 Toh Tuck Link, Singapore 596224
USA office: 27 Warren Street, Suite 401-402, Hackensack, NJ 07601

Library of Congress Cataloging-in-Publication Data
Names: Fanchi, John R., author.
Title: Confronting the enigma of time / John R. Fanchi, Texas Christian University, USA.
Description: New Jersey : World Scientific, [2023] | Includes bibliographical references and index.
Identifiers: LCCN 2022032911 | ISBN 9781800613188 (hardcover) |
ISBN 9781800613348 (paperback) | ISBN 9781800613195 (ebook) |
ISBN 9781800613201 (ebook other)
Subjects: LCSH: Space and time--History. | Time--History.
Classification: LCC QC173.59.S65 F36 2023 | DDC 530.11--dc23/eng20221013
LC record available at https://lccn.loc.gov/2022032911

British Library Cataloguing-in-Publication Data
A catalogue record for this book is available from the British Library.

For any available supplementary material, please visit
https://www.worldscientific.com/worldscibooks/10.1142/Q0389#t=suppl

Desk Editors: Jayanthi Muthuswamy/Adam Binnie/Shi Ying Koe

Typeset by Stallion Press
Email: enquiries@stallionpress.com

To my grandchildren Cameron, Caleb, and Luke

Preface

Confronting the Enigma of Time examines the role of time in modern physics. The enigma of time is a conceptual conflict between the two leading theories of modern physics: relativity and quantum theory. It raises the question of the fundamental nature of time in a physical sense. Attempts to unify the two theories have had limited success. We show that many of the issues which arise in understanding time occur when time is treated as if it is the same physical quantity in every application. The reader will gain an increased awareness of time and its place in our understanding of nature.

Supplementary references by chapter — denoted by superscripts in the text — are provided as Endnotes at the end of each chapter. Dates of Selected Characters presents a table of selected people, their countries of origin, and their dates of birth and death. The dates help establish a timeline for the material discussed in the book. References are located alphabetically by author in References, and are identified in the text by parentheses as in (author, date).

I would like to thank my IARD colleagues for insightful discussions over the years.

About the Author

John R. Fanchi has a Ph.D. in physics from the University of Houston and is the author of a variety of books in the areas of physics, earth science, mathematics, and engineering. He was co-founder and first President of the International Association for Relativistic Dynamics. He has worked in the energy industry and has taught courses in energy, engineering, and physics at Texas Christian University (TCU), Colorado School of Mines (CSM), and the University of Tulsa. His books include *Parametrized Relativistic Quantum Theory* (1993, Kluwer), *Shared Earth Modeling* (2002, Butterworth-Heinemann), *Energy: Technology and Directions for the Future* (2004, Elsevier Academic), *Math Refresher for Scientists and Engineers* (2006, 3rd edition, John Wiley and Sons), *Energy in the 21st Century* (2017, 4th edition, World Scientific), *Principles of Applied Reservoir Simulation* (2018, 4th edition, Elsevier), *The Goldilocks Policy: The Basis for a Grand Energy Bargain* (2019, World Scientific), and *Reason, Faith, and Purpose: The Ultimate Gamble* (2021, World Scientific).

Contents

Chapter 1

Introduction to the Enigma of Time

What is time? The meaning of time has changed over the millennia. Physicist Lee Smolin has called this question

> **Q1.1**. "the single most important problem facing science." (Smolin, 2013, p. xi)

Smolin said that we perceive life as a flow of moments and believed that

> **Q1.2**. "to make sense of the picture of the universe that cosmological observations are bringing to us, we must embrace the reality of time in a new way. This is what I mean by the rebirth of time." (Smolin, 2013, p. xii)

Author H.G. Wells introduced a Time Traveller in *The Time Machine* who believed that

> **Q1.3**. "there is no difference between Time and any of the three dimensions of Space except that our consciousness moves along it." (Wells, 1895, p. 4)

Theologian and philosopher Augustine of Hippo (354–430), also known as Saint Augustine, recognized the difficulty of defining time. Augustine asked

Q1.4. "what then is time? If no one asks me, I know what it is. If I wish to explain it to him who asks me, I do not know." (Augustine, ca. 398, Book XI, Chapter XIV)

Time appears in many of the equations of modern science. For example, the role of time was a key difference between the views of Isaac Newton and Albert Einstein. To Newton, time was a monotonically increasing 'arrow' that parametrized the direction of evolution of a system. Einstein rejected Newton's concept and identified time as the fourth coordinate of a spacetime four-vector. The question of the meaning of time depends on the resolution of many temporal issues, such as those illustrated in Table 1.1.

The first column in Table 1.1 recognizes that the concept of time has been viewed historically as an illusion, a reality, or an emergent concept. Some have viewed time as a figment of our imagination; others have said that time is as real as spatial coordinates; and still others consider time an emergent concept, that is, a concept that emerges from other, more fundamental, concepts. If time is real, what are its physical characteristics? Is there only one time, or is it possible that there is more than one concept that exhibits temporal characteristics as illustrated in the second column? Is there only one kind of time, or has the concept of time been used as a catch-all term for multiple concepts, each with distinguishable characteristics? Is time an absolute ordering parameter as suggested in the third column, or does the interpretation of time depend on the observer?

Today, scientists are considering different concepts of time as a means of resolving the problem of time, that is, the problem of resolving conceptual incompatibilities between the notions of time used in different physical disciplines, notably general relativity and quantum mechanics. Some scientists are trying to show that time is not real; that it is an emergent property of a system rather than a fundamental property. Others have

Table 1.1. Temporal Issues.

What Is Time?	How Many Times?	Time Is
Time is an illusion		
Time is real	One time	Absolute or relative
	Two times	Absolute and relative
Time is emergent		

hypothesized the need for two temporal variables: a coordinate time (Einsteinian time), and an evolution parameter (an invariant form of historical time). It is fair to say that time is an enigma.

# # #

We begin confronting the enigma of time by reviewing the history of time from prehistory to Isaac Newton's mechanics and James Clerk Maxwell's electromagnetic theory in Part 1. The role of time in nonrelativistic classical physics is discussed, and important temporal anomalies are identified.

The replacement of Newtonian time with the times of relativity, Minkowski time and Einsteinian time, is described in Part 2. This material extends our review of the history of time from Newton to Einstein.

The role of time in developing an understanding of cosmology and the evolution of the inflationary universe is discussed in Part 3. We must consider modern physical concepts from the subatomic scale to the cosmological scale. In the process, we introduce ideas from quantum field theory at the subatomic scale that are influencing the future of time discussed in Part 4.

Several questions about the future of the concept of time are considered in Part 4. They include:

- How do we explain the apparent flow of time in a single direction?
- Is the arrow of time necessary?
- Is time an illusion or is it real?
- Should time be considered an emergent property, or is it a fundamental property?
- Do we need two temporal variables to understand the physical universe at scales ranging from subatomic to cosmological?

Part 1

Newtonian Time

Chapter 2

Time from Prehistory to the Ancient Greeks

Human behavior and our understanding of time have been linked since the beginning of human culture. The role of time from prehistory to the ancient Greeks is reviewed here.

2.1 The Stone Age

Our understanding of time has evolved since human prehistory. The human family tree consisting of protohumans and humans is summarized in Table 2.1. The only surviving species of the genus *Homo* in the modern world is *Homo sapiens.*[1] In our discussion of time, human refers to *Homo sapiens*.

A history of early human development can be expressed in terms of three successive stages of technological progress[2]: the Stone Age, the Bronze Age, and the Iron Age. The boundary between the Stone Age and the Bronze Age is marked by the appearance of artifacts made from bronze, a metal composed of a mixture of copper and tin. The significance of the distinction between the Bronze Age and the Iron Age was the development of new ways to heat iron. Copper and tin have a lower melting point than iron. The ability to melt iron made it possible to craft tools and weapons with more strength and resilience than bronze materials.

The Stone Age is subdivided into the Paleolithic or Old Stone Age, the Mesolithic or Middle Stone Age, and the Neolithic or New Stone

Table 2.1. The Human Family Tree.

Group	Species	Years Before Present (YBP)* Lived from	To
Homo	*Homo sapiens*	300,000	Present
	Homo floresiensis	100,000	50,000
	Homo neanderthalensis	400,000	40,000
	Homo heidelbergensis	700,000	200,000
	Homo erectus	1.89 million	110,000
	Homo rudolfensis	1.9 million	1.8 million
	Homo habilis	2.4 million	1.4 million
Paranthropus	*Paranthropus robustus*	1.8 million	1.2 million
	Paranthropus boisei	2.3 million	1.2 million
	Paranthropus aethiopicus	2.7 million	2.3 million
Australopithecus	*Australopithecus garhi*	2.6 million	2.5 million
	Australopithecus africanus	3.3 million	2.1 million
	Australopithecus bahrelghazali	3.6 million	3 million
	Australopithecus afarensis	3.85 million	2.95 million
	Australopithecus anamensis	4.2 million	3.8 million
Ardipithecus	*Ardipithecus ramidus*	4.6 million	4.4 million
	Ardipithecus kadabba	5.8 million	5.2 million
	Orrorin tugenensis	6.2 million	5.8 million
	Sahelanthropus tchadensis	7 million	6 million

Note: *All YBP are approximate.

Age. The Paleolithic began approximately 2.6 million YBP (years before present). This corresponds to the age of the oldest-known stone tools. The Paleolithic can be subdivided into Lower, Middle, and Upper Paleolithic Periods based on technologies available during the period. Time boundaries between the Lower, Middle, and Upper Paleolithic Periods are loosely defined because technologies emerged at different times in different parts of the world. A Three-Age framework of prehistory is outlined in Table 2.2.

The Lower Paleolithic began approximately 2.6 million YBP and is associated with the emergence of chopping tools, hand axes, and cleaving

Table 2.2. Three-Age Framework of Prehistory.

Age	Period		Beginning (YBP*)
Stone	Paleolithic	Lower Paleolithic	ca. 2.6 million
		Middle Paleolithic	ca. 250,000
		Upper Paleolithic	ca. 40,000
	Mesolithic		ca. 12,000
	Neolithic		Varies
Bronze			ca. 5,500
Iron			ca. 3,200–2,600

Note: *YBP = years before present.

tools. The Middle Paleolithic began about 250,000 YBP. It is characterized by the widespread use of fire and the use of flake tools made from carefully shaped flakes of flint. The Upper Paleolithic covers the duration from approximately 40,000 YBP to 12,000 YBP.

Upper Paleolithic artifacts were more sophisticated than artifacts from the Lower Paleolithic and Middle Paleolithic. In addition to stone, Upper Paleolithic artifacts were made from organic materials such as antler, bone, leather, and wood. The discovered artifacts represented a variety of tool types and exhibited greater complexity and specialization than artifacts from earlier periods. Distinctive artistic traditions began to appear in different regions and included paintings, sculptures, and musical instruments. Artifacts found at different locations around the world and the behavior of contemporary hunter-gatherers have suggested that humans lived as hunter-gatherers during the Palaeolithic.

The end of the last Ice Age approximately 12,000 YBP and the beginning of an interglacial age mark the boundary between the Paleolithic and Mesolithic. The interglacial age was warmer than the ice age, and glaciers from the ice age began to retreat. The Mesolithic was populated by nomadic hunter-gatherers living in a warming world.

The beginning of the Neolithic varied from one part of the world to another. The boundary between the Mesolithic and Neolithic was marked by the appearance of agriculture. The Neolithic was characterized by the formation of enduring villages, cereal cultivation, animal domestication, and the development of crafts such as weaving and pottery. Stone tools were still used, but they were now refined by grinding and polishing.

Farming and the cultivation of cereal grains allowed formerly nomadic people to congregate in villages consisting of long-lasting structures.

Human perception of time changed significantly between the Paleolithic and Neolithic periods. Human culture changed from a dependence on hunter-gatherers to a dependence on farming. On a daily basis, hunter-gatherers in the Paleolithic would have noticed a change from day to night, while farmers had to manage daily activities. In *About Time*, astrophysicist Adam Frank observed that "the hunter-gatherer lived through time as an unbroken whole, the farmer lived within a time marked by the daily rounds of animal husbandry, home maintenance, and village life" (Frank, 2011, p. 23).

2.2 The Urban Revolution

The next major change in our interaction with time occurred during the Urban Revolution.[3] Australian archaeologist Verne Gordon Childe (1892–1957) introduced the term 'urban revolution' into archaeology. The term refers to the transition of small farming settlements into complex urban societies supported by extensive empires.

The earliest urban revolution is dated to approximately 5,000 YBP in the area bounded by the Tigris and Euphrates Rivers known as the Fertile Crescent. Sumerian society in Mesopotamia — from the Greek for 'between rivers' — emerged from the urban revolution. Sumerians were known for innovations in language and governance. Some Sumerians could read and write, and Sumerian art included mosaics and mural paintings. Sumerians built the ziggurat, a pyramid-like, stepped temple, and Sumerian architecture included terra cotta ornamentation with bronze accents. Sumer is considered the first civilization from a modern perspective.

Sumerian culture consisted of a group of city-states, including Kish, Ur, and Uruk. City-states were surrounded by walls with settlements outside of the walls. The population of Uruk at its peak ranged between 40,000 and 80,000 people within the city's walls. Moring said that *The Epic of Gilgamesh* was told on clay cuneiform tablets (Moring, 2002, p. 5). *The Epic of Gilgamesh* is the story of historical king Gilgamesh of Uruk in the Mesopotamian state of Babylonia. Gilgamesh lived during the second millennium BCE.

Frank wrote "that explicit and accurate calendars were required to meet Mesopotamia's agricultural, economic and political needs"

(Frank, 2011, p. 33). Calendars written on cuneiform tablets during the urban revolution subdivided the day into explicit divisions. These daily divisions helped people complete more specialized tasks. The formation of communities made it possible for people to have time to focus their activities on specialized activities.

2.3 The Ancient Greek View of Time

The Ionian influence

Ionia was a region of islands and coast along the eastern shore of the Aegean Sea and the western shore of Asia Minor, which is now the western coast of modern-day Turkey. Ionia was settled by the Greeks during the 11th century BCE following the collapse of the Hittite empire.[4] A confederation of Ionian city-states was formed by 8th century BCE. Ionian philosophy dominated the intellectual life of Greece during the period from 8th century BCE to 5th century BCE. The Ionian city of Miletus is considered the birthplace of natural philosophy, and Ionians helped establish the foundations of Greek philosophy.

Farmer-poet Hesiod (ca. 750–650 BCE) wrote the didactic poem *Works and Days* ca. 700 BCE. A didactic poem combines poetry with information or instruction. Hesiod tells the reader how to lead an orderly, structured life in the mountainous farming environment of the Peloponnese peninsula in southern Greece. His work demonstrated that the early Greeks thought time was a means of ordering observable natural events. It also displayed a relationship between Greek gods and everyday life.

Thales (ca. 624–546 BCE) was an Ionian philosopher who sought to explain the world without attributing every natural event to the supernatural. In 585 BCE, Thales was the protagonist in a story that said he used Egyptian geometry and Babylonian astronomy to prepare a mathematical model of the motion of the Sun and Moon to predict a solar eclipse. Frank said that "modern scholars believe this story may be more myth than reality…the mythologizing of Thales' mathematical feat by enthusiastic Greek philosophers tells us a great deal about the swift but seismic cultural swift that had occurred in the Hellenistic world" (Frank, 2011, p. 43). Ionian Greeks were beginning to rely on rational thought rather than theology to understand the natural world.

Thales' students continued his approach. According to Frank, Thales' pupil Anaximander (ca. 610–546 BCE) "proposed the world was built of 'intermingled opposites: hot and cold, dry and wet, light and dark'. The tension between these opposites produced a dynamic, evolving world. In Anaximander's account, all animals and humans evolved from lesser ocean creatures — a prototype of Charles Darwin's vision" (Frank, 2011, p. 43).

The Ionian Pythagoras (ca. 570–490 BCE) believed that mathematics was essential to a description of nature. American physicist Michio Kaku recounted a legend that said Pythagoras "noticed similarities between the sound of plucking a lyre string and the resonances made by hammering a metal bar. (The sounds) created musical frequencies that vibrated with certain ratios" (Kaku, 2021, p. 9).

Pythagoras established a school that became known as the Pythagorean school. The school developed many of the mathematical tools used by Greek philosophers. For example, the Pythagorean vision relied on the sphere and five symmetrical three-dimensional forms constructed using geometric rules. The three-dimensional forms were later called Platonic solids. Centuries later, Johannes Kepler (1571–1630) tried to model planetary orbits in the Solar System using the five Platonic solids and the heliocentric model proposed by Nicolaus Copernicus (1473–1543). Kepler replaced the Platonic solids with elliptical orbits when he realized the Platonic solids did not work.

The Pythagorean reliance on mathematics led Frank to conclude that, "For the Pythagoreans, reality was mathematics" (Frank, 2011, p. 44). By contrast, Heraclitus of Ephesus, Ionia (535–475 BCE), believed that change is the only reality. He interpreted the constant state of flux to mean that permanence in nature was unknowable. Heraclitus likened change to the flow of a river, that is, you cannot step twice into the same river.

The Eleatic view

Heraclitus' views were challenged by Parmenides (ca. 515 BCE–?) from Elea, a Greek colony in Italy. Parmenides believed that things cannot come from nothing or disappear into nothing. Movement of a thing means moving into a void where nothing existed before, but the void (nothing) cannot exist. Therefore, the appearance of change and the appearance of movement cannot be real. To find reality, we must find that which is permanent and unchanging. Anything that is changing is not permanent and cannot be real. Since time changes, time must be an illusion.

The Eleatic school of thought represented by Parmenides believed that things could exist and rejected the existence of change in the world. A middle ground between Heraclitus' view of perpetual change and the Eleatic view of permanent, unchanging reality was provided by the atomists.

The atomist view

The founding of atomism is attributed to Leucippus of Miletus (ca. 5th century BCE) and Democritus of Abdera of Trace (ca. 460–370 BCE). Abdera was east of Thessaloniki, Greece. Democritus wondered what would happen if you broke or cut an object into smaller and smaller pieces of matter. Eventually, you would obtain indivisible bits of matter called atoms, from the Greek adjective meaning 'uncuttable'. Democritus postulated that everything is made of atoms that are perpetually moving, physically indivisible objects which are real and permanent. Permanence and indivisibility are fundamental notions of atomism.

Greek atomism was a deterministic view of nature: there is a direct cause for everything that happens, including the motion of atoms. The atomists did not provide an explanation for the motion of atoms. Successors of the atomists, notably Plato (428–347 BCE) and Aristotle (ca. 384–322 BCE), sought to provide an explanation.

Plato

Plato (428–347 BCE) believed that a specific thing may change or disappear, but the Idea or Form embodied by the thing is unchanging and real. Like the Pythagoreans, Plato argued that Ideas can be known only by reason, such as mathematical proofs, while observations are merely sense experiences. He introduced the allegory of the cave to illustrate his view that the world we perceive is merely a projection of reality.

Socrates (ca. 470–399 BCE) presented Plato's allegory of the cave in Book VII of The Republic.[5] In the allegory of the cave, prisoners lived their entire lives chained in a cave. They can only see the cave wall in front of them and cannot see behind them. They do not know about a fire burning behind them or the world outside of the cave.

A barrier stands between the prisoners and the fire. Unbeknownst to the prisoners, puppeteers can cast puppet shadows on a cave wall facing the prisoners. The prisoners can see the puppet shadows but are unaware that the shadows are being cast by real puppets. The prisoners name the shadows and mistakenly talk about the shadows as real objects. Their language confuses shadows with the reality of real objects.

When the prisoners are released, one of the prisoners leaves the cave and sees the real forms of objects in the sunlight. The freed prisoner realizes that the shadows he saw in the cave were not real. He returns to the cave and tries to explain what he saw to the prisoners remaining in the cave. They do not believe him.

The Ideas or Forms of objects were Plato's raw material. The source of reality of the Ideas is the ultimate Idea which Plato called the Good. The Good was the ultimate reality and source of all that was knowable in the real world. Christian Platonism equates Plato's Good with God.

One of Plato's students, Eudoxus of Knidus in Asia Minor (ca. 408–355 BCE), developed a geometrical, Earth-centered (geocentric) model of the cosmos. In his model, Earth was surrounded by a concentric set of rotating spherical shells. Each of the visible planets, Moon, and Sun was attached to its own spherical shell. Eudoxus tuned the rotation of each shell to replicate contemporary observations.

Plato's view of time was shaped by his belief that human beings occupy both the world of sense experiences and the real world of Ideas. He argued that "the time-bound world we experience so vividly is a corrupted version of the ideal and timeless world of mathematical forms" (Frank, 2011, p. 46). American physicist Lee Smolin said the two worlds are "a world bound in time and a timeless world" (Smolin, 2013, p. 9). From this point of view, Plato's world of ideas exists outside of time.

Aristotle

Plato's student Aristotle (ca. 384–322 BCE) played a significant role in history. Aristotle was Alexander the Great's tutor in Macedonia while Alexander was a youth. Years later, when he was 50 years old, Aristotle founded his school, the Lyceum, near Athens, while Alexander was conquering Asia (Amadio and Kenny, 2021). Aristotle made many contributions to a range of subjects, including metaphysics, cosmology, and physics.

In metaphysics, Aristotle believed we could use sense experience to understand nature. By contrast, his teacher Plato believed reason should hold a primary role in metaphysics and rejected sense experience.

Aristotle adopted the more pragmatic view of elevating sense experience to the primary role. He argued that we could begin understanding nature by recognizing that all objects are changing, and then identify that which changes from that which remains the same. That which is unchanging is real. Thus, Aristotle concluded that motionless objects are the basic building blocks of the universe.

Furthermore, the goal of every object was to attain rest and unchangeability, a state he called the 'prime mover'. Aristotle's Prime Mover, which is also known as Unmoved Mover, is a being that does not act, yet attracts everything else by its presence. It is the Aristotelian God.

In physics, Aristotle's view of nature[6] was documented in *Physics, On the Heavens*, and *On Generation and Corruption*. *Physics* (Aristotle, 350 BCE) was adopted as a scientific text of the Medieval Catholic Church. Aristotle rejected atomism because he did not believe human beings were composed of inanimate objects without souls. Instead, Aristotle introduced five basic elements: earth, water, fire, air, and ether.

In cosmology, Aristotle adopted a geocentric model in which he divided the world into an earthly realm and a celestial realm. The natural motion of earth and water is toward the center of the world, while the natural motion of fire and air is toward the celestial realm. Heavenly bodies (the Moon, Sun, and visible planets) are made up of a fifth substance called ether, and the natural motion of the ether is circular around the world.

Aristotle considered spatial extension, motion, and time as continua that exist in ordered relation to one another. Motion depends on spatial extension, and time depends on motion. In this view, we observe how much time passes by observing motion, or the process of change relative to a spatial continuum. Time does not pass if there is no change.

2.4 The Hellenistic Period

Alexander III of Macedon (356–323 BCE), also known as Alexander the Great, became King of the Greek Kingdom of Macedonia in 336 BCE. He assumed the throne when he was 20 years old after his father Philip was assassinated. Alexander eliminated his rivals and began an empire-building campaign that lasted until his death in 323 BCE. Alexander's cause of death is unknown. It may have been due to malaria, a natural cause, or poison.

Table 2.3. The Hellenistic Period in History.

Event	Date
Alexander the Great becomes King	336 BCE
The death of Alexander the Great	323 BCE
Hellenistic Period	323–331 BCE
Rome conquers the Hellenistic territories	31 BCE

Table 2.4. Ancient Greek View of Time.

Person	Period	View
Hesiod	ca. 750–650 BCE	Time was a means of ordering events
Thales	ca. 624–546 BCE	Rely on rational thought, not theology
Anaximander	ca. 610–546 BCE	The world is dynamic and evolving
Heraclitus	535–475 BCE	Change is the only reality
Parmenides	ca. 515 BCE–?	Time is illusory
Democritus	ca. 460–370 BCE	Atoms are permanent
Plato	428–347 BCE	The world of ideas is timeless
Aristotle	ca. 384–322 BCE	Time depends on change

Alexander the Great's empire spread Greek ideas and culture from the Eastern Mediterranean to Asia. It was the largest empire ever seen at the time. The period between the death of Alexander the Great and the Roman conquest of the territories once ruled by Alexander lasted from 323 BCE to 31 BCE. The era is known as the Hellenistic Period and is displayed in Table 2.3. The adjective 'Hellenistic' refers to the approximately three centuries of Greek culture and influence following the death of Alexander.

# # #

The ancient Greek view of time through the atomists is summarized in Table 2.4. A new vision of nature, order, and time spread during the Hellenistic Period. The vision was based on Greek ideals in philosophy and science.

Endnotes

1. The evolution of humans and protohumans is discussed by Tudge (2000), Pickrell (2006), Coward (2009), Palmer (2009), Teerikorpi *et al.* (2019), and the Smithsonian Institution (2020).
2. Stone Age time frames are discussed by Violatti (2014) and Keesing *et al.* (2020).
3. The urban revolution is discussed by Moring (2002), Frank (2011), Britannica (2016), and History.com Editors (2021).
4. The Ancient Greeks are discussed by Moring (2002), Frank (2011), Britannica (2012), History.com Editors (2019, 2020), Hesiod (ca. 700 BCE), and Teerikorpi *et al.* (2019).
5. Sources on Plato include Cronk *et al.* (2004, Book VII, pp. 118–123) and Plato (1991).
6. Sources on Aristotle include Barnes (1991), Rovelli (2015), and Amadio and Kenny (2021).

Chapter 3

Time from Ptolemy to Kepler

Augustine of Hippo (354–430) considered questions about the biblical meaning of time and creation in *The Confessions of Saint Augustine*, published around 398. Referring to the Christian God, Augustine argued that "thou art the creator of all times… In the eminence of thy ever-present eternity, thou precedest all times past, and extendest beyond all future times" (Augustine, 398, Book XI, Chapter XIII). To Augustine, God created time and transcends time.

Augustine believed that God was eternal: "In eternity, God is before all things" (Augustine, 398, Book XII, Chapter XXIX). He reasoned that God first made unformed matter as a primal formlessness. Heaven, Earth, and time emerged from this primal formlessness: "the matter of things was first made and was called 'heaven and earth' because out of it the heaven and earth were made. This primal formlessness was not made first in time, because the form of things give rise to time" (Augustine, 398, Book XII, Chapter XXIX).

Augustine's philosophy was influenced by the ancient Greeks, notably Plato. Augustine adopted the geocentric view of heaven and earth and believed in a creator who created heaven, earth, and time. His views helped shape the doctrines of the Roman Catholic Church, but those doctrines were challenged during the historical period covered here.

3.1 Ptolemaic System or Copernican System?

Major contributors to our understanding of the Solar System during the period from Aristarchus of Samos to Galileo are listed in Table 3.1.

Table 3.1. Contributors to the Modern View of the Solar System.

Person	Period	Contribution
Aristarchus of Samos	ca. 310–230 BCE	Heliocentric hypothesis
Ptolemy	ca. 100–170	Geocentric hypothesis
Nicolas Copernicus	1473–1543	Revived heliocentric hypothesis
Tycho Brahe	1546–1601	Accurate observation of orbits
Johannes Kepler	1571–1630	Three laws of planetary motion
Galileo Galilei	1564–1642	Observe orbits with telescope

This period saw the development of ideas designed to model the motion of the Sun, Moon, and visible planets of our Solar System. Emerging from this period was the replacement of the geocentric model by the heliocentric model.

Aristarchus of Samos (ca. 310–230 BCE) is credited with hypothesizing the first Sun-centered, or heliocentric, model ca. 280 BCE. He observed the size of Earth's shadow on the Moon during a lunar eclipse. The observation helped Aristarchus formulate a model in which Earth and visible planets orbited a centrally located Sun.

Rather than applying a heliocentric model, astronomer Claudius Ptolemy (100–170) from Alexandria, Egypt, developed a geocentric model of celestial motion (Jones, 2020). He was able to match available data, including retrograde motion in which the body seemed to reverse the direction of its orbit, by replacing simple geometric motions with a combination of motions (Figure 3.1). Ptolemy modeled the motion of heavenly bodies around the stationary Earth using geometrically perfect circles. The apogee and perigee of an orbiting body were modeled with a great circle called the deferent, and retrograde motion was modeled by adding a smaller circle called an epicycle. The center of the epicycle moves on the circumference of the larger circle.

Ptolemy's geocentric model was widely accepted for over 1,000 years until Polish astronomer and priest Nicolas Copernicus (1473–1543) replaced Ptolemy's geocentric model with a heliocentric model in 1515. Copernicus considered Earth to be a planet like the other known planets. The planets in the Copernican theory were Mercury, Venus, Earth, Mars, Jupiter, and Saturn. All planets, including the Earth, orbited the Sun. The Moon orbited Earth while Earth orbited the Sun. Concerned about the reaction of the Roman Catholic Church to his heliocentric theory,

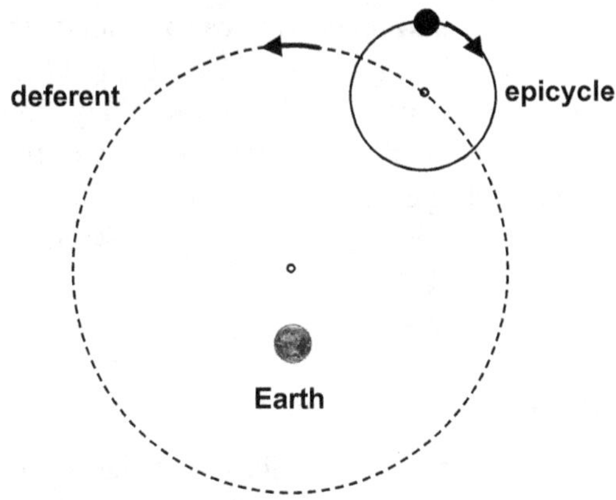

Figure 3.1. The Ptolemaic System.

Copernicus did not publish his theory until 1543, the year of his death. He published his findings in *De Revolutionibus Orbium Coelestium* (*On the Revolutions of the Heavenly Spheres*) (Copernicus, 1543).

Danish astronomer Tycho Brahe (1546–1601) accurately recorded the positions of stars and planets. He used a modified geocentric model with Earth at the center and the Sun orbiting Earth. In this model, which is also known as the Tychonian model, the Moon orbited Earth, and the planets orbited the Sun. Brahe and his assistants used naked-eye astronomical observations to accumulate "the best measurements of planetary motions that had ever been made. These sat in his record books undecoded until 1600 when he employed…Johannes Kepler" (Smolin, 2013, p. 16).

German astronomer Johannes Kepler (1571–1630) replaced Brahe's geocentric model with a heliocentric model to develop a mathematical model of planetary motion in the Solar System. His mathematical model represented planetary orbits as ellipses rather than circles. The resulting Sun-centered model consisted of three laws that Kepler corroborated using Brahe's measurements.

Kepler abandoned the Ptolemaic system favored by the Roman Catholic Church when he adopted the Copernican system. By contrast, he accepted the Church's view of time when he used the Bible to calculate "the exact date of Genesis, arriving at the date 3992 BCE for the creation of the world" (Frank, 2011, p. 83). For comparison, biblical scholar James Ussher (1581–1656) of Ireland, Archbishop of Armagh, used biblical family histories in Genesis and other sources to estimate the date of creation as October 22, 4004 BCE (Hart-Davis, 2011, p. 68). Astrophysicist Adam Frank pointed out that "Ussher's chronology drew on Persian, Greek, and Roman sources in order to create a history of the ancient world that remains in remarkable agreement with modern accounts. It was the truly ancient history, of course, that Ussher got wrong" (Frank, 2011, p. 84; see also Bergman, 2014).

3.2 Galileo Galilei

An instrument that could provide more information than the naked-eye astronomical observations recorded by Brahe appeared in 1608. The American Institute of Physics (AIP) credited Hans Lippershey (ca. 1570–1619), a Dutch eyeglass maker, with inventing a telescope in 1608:

> **Q3.1**. "The telescope first appeared in the Netherlands. In October 1608, the national government in The Hague discussed a patent application for a device that aided 'seeing faraway things as though nearby.' It consisted of a convex and concave lens in a tube. The combination magnified objects three or four times. The government found the device too easy to copy and did not award a patent, but it voted a small award to Jacob Metius and employed Hans Lippershey to make several binocular versions, for which he was well paid." (AIP Telescopes, 2021)

News of the telescope spread rapidly throughout Europe. Approximately a year after the invention of the telescope, Italian scientist Galileo Galilei (1564–1642) designed a telescope that made it possible for him to make several important observations.[1] Galileo made sketches of his telescopic observations, which included four moons of Jupiter orbiting the planet Jupiter, Saturn's rings, mountains and craters on an imperfect surface of the Moon, sunspots, and the phases of Venus. In 1610, Galileo published a small book that presented some of his work in *Sidereus Nuncius* — translated as *The Sidereal Messenger*, which is also known as *The Starry Messenger*.

Galileo realized that he could explain planetary motions if he adopted the heliocentric hypothesis of Copernicus. For example, if he assumed Earth and other planets orbited the Sun, they would exhibit orbital motion similar to the orbits of Jupiter's moons around Jupiter. This brought him to the attention of the Roman Catholic Church.

Unlike his contemporaries, Galileo chose to publish his findings in Italian rather than the more traditional Latin or Greek. As a consequence, Galileo's work with the heliocentric model could be widely read and understood by the public. This created a problem for the Roman Catholic Church, which considered the heliocentric model heresy.

The Church banned Galileo from publishing more material about the heliocentric model in 1616. Rather than complying, Galileo gathered more supporting evidence and published a comparison of the Church-sanctioned geocentric model (Ptolemaic system) with the heliocentric model (Copernican system) in *Dialogue Concerning the Two Chief World Systems: Ptolemaic and Copernican* (Galilei, 1632). The Inquisition of the Roman Catholic Church responded in 1663 by forcing Galileo to recant his beliefs and placing him under house arrest. Centuries later, in 1992, Pope John Paul II reversed the official position of the Roman Catholic Church. The Pope declared that Galileo was right based on the findings of a committee of the Pontifical Academy. (SPACE Editors, 1992).

Table 3.2 summarizes Galileo's major contributions to our understanding of nature. In addition to his work in astronomy, Galileo made contributions to physics and mathematics. In *Il Saggiatorore* (*The Assayer*), Galileo discussed his ideas of physical reality and a scientific method based on observation and the use of mathematics as the language of science. In *Dialogues Concerning Two New Sciences*, Galileo reported his findings on the strength of materials, notably the bending and breaking of columns, and the motion of objects. He used algebra to show that the trajectory of objects moving with constant acceleration could be mathematically described by a parabola.

Table 3.2. Galilean Contributions to a More Modern View of Nature.

Publication	Date	Topic
The Sidereal Messenger	1610	Observations with a telescope
The Assayer	1623	New scientific method
Dialogue Concerning the Two Chief World Systems	1632	Compare geocentric and heliocentric systems
Dialogues Concerning Two New Sciences	1638	Experimental study of materials and trajectories
Du Moto (*On Motion*)	1687	Challenges Aristotelian physics

The last publication shown in Table 3.2, *Du Moto* (*On Motion*), was published after Galileo's death. It included content that was largely written while Galileo was teaching mathematics at the University of Pisa in the period from 1589 to 1592. His findings in *The Assayer*, *Dialogues Concerning Two New Sciences*, and *On Motion* contradicted some of Aristotelean physics. For example, Aristotle believed that the motion of a freely falling body is proportional to its weight. By contrast, Galileo found that the motion of freely falling bodies does not depend on mass. Galileo's results disputed the widely accepted science of Aristotelian physics, which did not please some of his colleagues. Galileo's contract was not renewed, but his patrons helped secure him a new position at the University of Padua.

Galileo's pendulum clock

Early in his career, while a student of medicine, Galileo observed the behavior of a lamp swinging in the Duomo of Pisa (Mitchell, 2013; Hart-Davis, 2011). The swinging mass behaved like a simple pendulum — a swinging mass attached to a string. Galileo used his pulse as a measurable, reproducible, and periodic interval to measure the duration of a pendulum swing. He observed that the duration of a pendulum swing did not depend on how far the pendulum swung or the mass of the pendulum. The duration of the pendulum swing did depend on the length of the pendulum string: the pendulum moves more slowly as the length of the string increases.

Galileo built a small portable machine called a pulsilogon, or pulse meter, to measure short time intervals based on his knowledge of the pendulum. The pulsilogon consisted of a pendulum supported by a wooden stand. He calibrated the pulsilogon with the human pulse and used it to measure the pulses of his patients.

Years later, near the end of his life and under house arrest, Galileo realized that the regularity of the pendulum swing made the pendulum suitable for regulating a mechanical clock. Galileo shared this insight with his son Vincenzio and others, but Galileo did not live long enough to build the clock. In 1647, after Galileo's death, Grand Duke Ferdinand II ordered the construction of a pendulum clock based on Galileo's design. Georg Lederle of Augsburg built Galileo's pendulum regulator and installed it in the tower of Palazzo Vecchio, Florence (Newton, 2004, pp. 51–53).

Dutch scientist and inventor Christiaan Huygens (1629–1695) was inspired by Galileo's work and is credited with patenting the first pendulum clock in 1657 (Hart-Davis, 2011, pp. 182–185; Huygens Patent, 2017). It is worth noting here that mechanical clocks were being built to help regulate life in the 17th century.

3.3 Normal Science and Scientific Paradigms

The use of observation to test scientific hypotheses is an essential aspect of modern science. Karl Popper (1902–1994), an Austrian and British philosopher of science, said that knowledge is gained when scientific hypotheses are falsified. According to Popper, if a scientific hypothesis is not falsified, it should be considered corroborated rather than verified or proved. Popper's view of scientific knowledge was challenged in 1962 by American philosopher of science Thomas Kuhn (1922–1996) in *The Structure of Scientific Revolutions* (Kuhn, 1970).

The replacement of the geocentric model with the heliocentric model is an example of a scientific revolution. It represented a major change in our view of nature. Kuhn sought to explain the stages of a scientific revolution. A flow chart illustrating key features of Kuhn's theory is displayed in Figure 3.2.

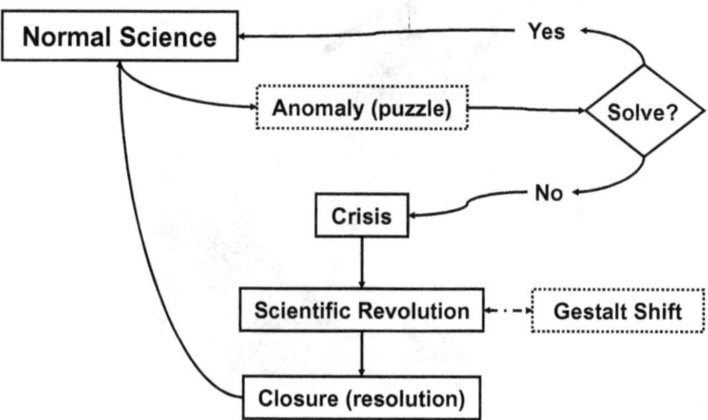

Figure 3.2. Flow Chart of Kuhn's Theory of Scientific Revolutions.

The term normal science in the upper left of Figure 3.2 refers to the mainstream scientific worldview or mainstream paradigm. An anomaly is a difficulty with normal science that can lead to a crisis in the practice of normal science if the anomaly cannot be resolved. In some cases, the anomaly can only be resolved and the crisis averted by changing our understanding or perception of normal change. This change is known as a gestalt shift.

Gestalt shifts require a change in perception. Figure 3.3 illustrates the concept of gestalt shift. The image shows both an old lady and a young woman, depending on your perspective.

Competing paradigms arise when the paradigm of normal science fails to explain important anomalies. Anomalies are viewed by advocates of the paradigm of normal science as puzzles to be solved within the context of normal science. The same anomalies are viewed as evidence of a crisis in normal science by challengers of the paradigm of normal science.

It may not be possible to rationally discuss conceptual differences or compare experimental results with colleagues during periods of crisis when competing paradigms are incommensurable, that is, they lack a common quality or measure for comparison. For example, competing

Figure 3.3. "My Wife and My Mother-in-Law" Was Drawn by Cartoonist W.E. Hill in 1915. Can You Make the Gestalt Shift?

paradigms can be incommensurable if they are expressed in a common language but the meaning of terms depends on your choice of paradigm. If competing paradigms are incommensurable, how does a new paradigm replace an old paradigm?

Kuhn argued that there can be many reasons why a new paradigm is adopted. Some reasons may be rational; some may be irrational. A proponent of the old paradigm may reasonably choose to adopt a new paradigm that is successful in resolving persistent anomalies associated with the old paradigm. Other people may abandon the old paradigm in favor of a new paradigm because of peer pressure or the loss of funding for the old paradigm. Kuhn did not support the idea that a new paradigm became the basis of normal science because the old paradigm was falsified.

If a new paradigm resolves the anomalies and results in the replacement of the current paradigm, closure is achieved, the period of crisis ends, and the new paradigm becomes the new normal science. The flow chart in Figure 3.2 is complete.

Falsification

Hungarian philosopher Imre Lakatos (1922–1974) attempted to support Popper's notion of falsification by arguing that scientists compare research programs instead of paradigms. A research program includes a hard core and a heuristic part. The hard core consists of a set of basic assumptions and perceptions. The heuristic part contains auxiliary hypotheses that help the practitioner conduct research. The research program may also embody widely accepted theories that are not yet unfalsified.

Lakatos said that normal science exists as long as the associated research program can make testable predictions in existing and new areas. The research program becomes embattled if new auxiliary hypotheses must be regularly adopted to resolve persistent anomalies or explain unexpected observations arising from competing research programs. An embattled research program is considered stagnating or regressive, while a successful research program is considered progressive.

There is significant overlap between Lakatos' research program and Kuhn's paradigm of normal science. A distinguishing characteristic of Lakatos' model of scientific revolutions when compared to Kuhn's is that

Lakatos said that scientists can use rational considerations or decisive experimental tests to justify replacing one research program with another.

# # #

The period from Aristarchus of Samos to Galileo saw the occurrence of two significant paradigm shifts. The first paradigm shift occurred when the Earth-centered Ptolemaic system was replaced by a Sun-centered Copernican system. This implied that Earth was no longer the center of creation. Indeed, it was now possible that there might not be a center of creation. From a scientific perspective, physicist Lee Smolin concluded that the failure of the Ptolemaic system could be considered a lesson that "neither mathematical beauty nor agreement with experiment can guarantee that the ideas a theory is based on bear the slightest relation to reality" (Smolin, 2013, p. 17).

A second paradigm shift occurred when Aristotle's hands-off approach to physics was replaced by direct observation of nature. Models of planetary motion were corroborated by comparing the models to observations. In his later years, Galileo provided another example of using experiments to corroborate his hypotheses by developing experimental methods for directly observing nature. The stage was set for the appearance of classical physics.

Endnote

1. Sources on the life and work of Galileo Galilei include Newton (2004), Aughton (2008), Hart-Davis (2011), Teerikorpi *et al.* (2019), and Helden (2021).

Chapter 4

Newton and Maxwell

Many of the natural systems encountered in everyday life are described by classical physics.[1] Systems that can be observed with our unaided senses are often referred to as macroscopic systems. For comparison, microscopic systems refer to systems that can only be seen with devices that aid our senses such as a microscope.

Several devices for measuring time intervals have been available for centuries. For example, the shadow cast by a sundial was used to indicate the time of day. A time interval could be measured with an hourglass. A typical hourglass consists of two identical bulbs connected by a narrow throat. The throat regulates the flow of a substance from the upper bulb to the lower bulb and can be used to calibrate a time interval. In a similar manner, a water clock can provide a steady stream of water for measuring time intervals by regulating the flow of water from a hole in the base of a container.

We saw in Chapter 3 that Galileo developed a pulsilogon, or pulse meter, to measure short time intervals when he was a student of medicine (Mitchell, 2013; Hart-Davis, 2011). Later in life, he performed experiments based on his hypothesis that the speed of a falling object would increase at a constant rate, or accelerate, as the object fell. An experimental study of a free-falling object could have provided insight into the validity of the Aristotelian view of motion, but the study of a free-falling object was hindered because devices for measuring time intervals associated with the high speeds attained by a free-falling object were inadequate. To solve this problem, Galileo reasoned that a free-falling object was equivalent to a ball rolling down a vertical plane. This suggested that the rate of change of speed of the ball could be reduced to measurable intervals by rolling the ball down an inclined plane.

Galileo used a water clock to study the motion of a small metal ball rolling down a linear groove in an inclined plane. His water clock was described in *Dialogue Concerning Two New Sciences*: "For the measurement of time, we employed a large vessel of water placed in an elevated position; to the bottom of this vessel was soldered a pipe of small diameter giving a thin jet of water, which we collected in a small glass during the time of each descent, whether for the whole length of the channel or for a part of its length; the water thus collected was weighed, after each descent, on a very accurate balance; the difference and ratios of these weights gave us the differences and ratios of the times" (Galilei, 1638, p. 179).

Galileo compared concepts of Aristotelian physics with his observations to determine their validity. According to physicist Carlo Rovelli, Aristotelian physics did not "properly characterize continuous acceleration…it was Galileo's triumph to understand first empirically (with the incline experiments) and then conceptually the central importance of acceleration. In this way, Galileo opened the way to [Isaac] Newton's major achievement on which modern physics is built: the main law of motion" that force is mass times acceleration (Rovelli, 2015, p. 34).

4.1 Newtonian Mechanics

Isaac Newton (1643–1727) wrote the *Principia*, or *Mathematical Principles of Natural Philosophy* (Newton, 1687). It became the foundation of classical physics, but it was almost not published. Earlier in his career, Newton was involved in a priority dispute with German mathematician Gottfried Wilhelm Leibniz (1646–1716) about their respective contributions to calculus. Newton's work in optics was later criticized by fellow Englishman Robert Hooke (1635–1703). Hooke had established a reputation as a pioneer in the use of microscopes and telescopes, and had coined the term 'cell'. Newton was reluctant to get involved in another controversy. In addition, the Royal Society of England, of which Newton was a member, could not afford to publish the *Principia*. Funding was provided by English astronomer and mathematician Edmund Halley (1656–1742). Halley was Astronomer Royal in Britain and is famous for calculating the orbit of a comet that later bore his name, Halley's comet.

He encouraged Newton to publish the *Principia* and provided funds for publication.

The *Principia* documented Newton's theory of the dynamics of macroscopic objects which is referred to by such names as Newtonian mechanics and classical mechanics. Newton hypothesized that space and time had the following characteristics: "Absolute, true, and mathematical time, of itself, and from its own nature, flows equably without relation to anything external… Absolute space, in its own nature, without relation to anything external, remains always similar and immovable" (Newton, 1687, p. 6). Unlike Aristotle, who believed that time does not pass if there is no change, Newton said that time continues its uniform flow even if nothing else changes. Similarly, space exists even if there is nothing there. Flowing time did not depend on the three immovable space dimensions. To Newton, absolute space and absolute time served as a mathematical framework for defining terms, expressing laws, posing problems, and presenting solutions. Measuring sticks calibrated space, and clocks measured the flow of time. Historian of science James Gleick noted that the notion of time "became more concrete when telegraphs were used to coordinate railroad schedules" in the 19th century (Gleick, 2016, p. 12).

Newton formulated three laws to describe the motion of objects in absolute space and absolute time (Newton, 1687, p. 13):

- Law I: "Every body continues in its state of rest, or of uniform motion in a right line, unless it is compelled to change that state by forces impressed upon it."
- Law II: "The change of motion is proportional to the motive force impressed; and is made in the direction of the right line in which that force is impressed…"
- Law III: "To every action there is always opposed an equal reaction…"

Aristotle's belief that every object seeks rest and unchangeability is contradicted by Newton's first law. The first law states that the motion of an object will be uniform and non-accelerating unless the object is acted upon by an external force. Newton's second law says that the change in momentum of an object during a given time interval depends on the forces acting on the object. The third law states that for every action by one object on another, there is an equal and opposite reaction.

Newton's theory is also known as Newtonian mechanics because of its focus on the motion of objects. It consists of absolute space, absolute time, and three laws of motion. In principle, we can calculate the historic and future behavior of an object from a set of initial conditions if we know all of the forces acting on the object during the period of interest. The initial conditions of the object are given by certain knowledge of its position and momentum at a given moment of time. A system is specified by assigning positions and momenta to each part of the system. The state of a system is defined by the position and momentum of each part of the system. The motion of the system, including its constituent parts, in space and time, is determined by solving Newton's laws of motion.

French scholar Pierre-Simon Laplace (1749–1827) "is usually credited with first clearly postulating scientific determinism: Given the state of the universe at one time, a complete set of laws fully determines both the future and the past" (Hawking and Mlodinow, 2010, p. 30). There is a direct cause for everything that happens in a deterministic universe once initial conditions are specified.

4.2 Causality

Many philosophers argued that Newtonian mechanics implied the universe was deterministic before the emergence of quantum theory early in the 20th century. The principles of quantum theory raised questions about the meaning of causality and determinism. Philosophical views of causality and determinism before the emergence of quantum theory are considered here.

French philosopher Rene Descartes (1596–1650), a predecessor of Newton, believed that causality existed in nature. To Descartes, three realities existed in the world: God, mind, and matter. Mind is the substance that thinks, and matter is a substance with extension. Mind and matter were created and preserved by a self-caused God.

Descartes was a dualist because he thought of mind and matter as two different substances. He also believed his extended objects — matter — could interact with one another and be understood in terms of mathematical properties. In Descartes' mechanistic view, extended objects and their interactions operated like a machine that was activated and maintained by God. If we knew and could apply God's natural laws, we could completely determine the future behavior of matter from past behavior. According to Descartes, God's role in a deterministic universe was to establish the initial conditions that set the universe in motion and then leave it alone.

Descartes' view of three realities was challenged by some of his contemporaries. For example, English philosopher Thomas Hobbes (1588–1609) was a monist: he said the only substance in the world was matter. The mind was a special type of matter. Hobbes' world was composed only of God and matter. His monistic view, or monism, is an example of a materialistic, deterministic view of nature.

Dutch philosopher of Portuguese origin, Baruch Spinoza (1632–1677), was neither a dualist like Descartes nor a monist like Hobbes. To Spinoza, all of reality was a single substance which could be understood rationally, and that substance was God. Spinoza did not distinguish between the creator and its creations. He believed that God and nature were inseparable.

Spinoza's god was not the God of his church, the Roman Catholic Church. The Church's explanation of nature was not necessary to Spinoza because everything in nature was just a different manifestation of God. The Church excommunicated Spinoza for advocating atheism. Nevertheless, Spinoza continued his scholarly work while earning a living as a lens grinder.

German mathematician and philosopher Baron Gottfried Wilhelm von Leibniz (1646–1716) developed a principle of sufficient reason based on his worship of a rational God. He believed that a perfectly rational God would have a reason for everything in God's creation. This implied to Leibniz that Newton's absolute time had no meaning because there was no reason "to prefer the universe to start at one absolute time rather than another" (Smolin, 2013, p. 28).

Leibniz introduced the monad as the basic entity in nature. His monad had no extension and did not interact with other monads. The motion of a monad did not depend on the motion of any other monad. The harmonious dance of monads is an example of a network of relationships. Leibniz envisioned a world enmeshed in a network of relationships. The world is a relational world where relationships precede space and can be used to define space.

Leibniz's view raised an important question: If monads are completely independent, how do we explain relationships that appear to be causal? Leibniz said that every monad must move in a pre-established harmony relative to every other monad. The harmony is orchestrated by God without contradiction because an all-wise, all-good, and all-knowing God selected the most harmonious of all possible worlds: "Through this means [selection] has been obtained the greatest possible variety, together with the greatest order that may be: that is to say, through this means has been obtained the greatest possible perfection" (Stroll and Popkins, 1972, p. 265).

Leibniz's beliefs contradicted the principles of Newtonian mechanics, but are present in some modern physical theories. For example, German physicist Albert Einstein (1879–1955) embraced Leibniz's relational view of space and time. Another example is the possibility of the existence of other worlds inherent in Leibniz's view. The Many-Worlds Interpretation (MWI) proposed by American physicist Hugh Everett III (1930–1982) in 1957 is one example. Everett hypothesized that all possible outcomes of a quantum measurement are physically realized in other equally real worlds or universes (Byrne, 2007). The set of universes is called the multiverse.

Scottish philosopher David Hume (1711–1776) did not agree with Descartes that the truth of all knowledge is only known through the mental world. Hume said that sense experience and empirical evidence must be used to determine the validity of theories. He expressed his skepticism at the end of *Enquiry Concerning Human Understanding*:

> **Q4.1**. "If we take in our hand any volume — of divinity or school metaphysics, for instance — let us ask: *Does it contain any abstract reasoning concerning quantity or number?* No. *Does it contain any experimental reasoning concerning matter of fact and existence?* No. Then throw it in the flames, for it can contain nothing but sophistry and illusion." (Hume, 1748, Section XII, Part III)

According to Hume, "...all our reasonings concerning causes and effects are derived from nothing but custom; and that belief is more properly an act of the sensitive, than of the cogitative part of our nature" (Russell, 1972, p. 671). Newton's laws are causal relationships based on empirical evidence. A causal relationship exists between two events A and B when the appearance of event A is always accompanied by the appearance of event B. For example, a bat strikes a ball, and the motion of the ball changes. We can hypothesize a causal relationship that says the motion of the ball was changed when the ball was struck by the bat. Once a causal relationship is specified, is it reasonable to assume that the causal relationship will recur in the future? As a rule, causal relationships were directed from the past to the future. If our historical sense experience tells us that event B always follows event A, it is reasonable to hypothesize that event A caused event B.

Hume recognized that causal relationships could be broken. When a causal relationship is broken, or falsified, we encounter an anomaly. The anomaly may be evidence of an underlying reality that has yet to be discovered.

German philosopher Immanuel Kant (1724–1804) believed that Hume relied too heavily on sense experience. Kant sought knowledge that was necessarily and universally true. Some of this knowledge should apply to all possible experiences and could be used to anticipate, or predict, more.

Kant was concerned about Hume's understanding of necessary (indisputable) knowledge and the origin of necessary knowledge. According to Kant, reason and sense experience were complementary:

> **Q4.2**. "The question was not whether the concept of cause was right, useful, and even indispensable for our knowledge, for this Hume never doubted; but whether that concept could be thought by reason **a priori**, and consequently whether it possessed an inner truth, independent of all experience, implying a wider application than merely to the objects of experience. This was Hume's problem. It was a question concerning the **origin**, not concerning the **indispensable need** of the concept." (Stroll and Popkins, 1972, p. 272)

Kant suggested that the origin, or source, of necessary knowledge was not the observed events, but the observer of the events:

> **Q4.3**. "I therefore first tried whether Hume's objection [to **a priori** causality] could not be put into a general form, and soon found that the concept of the connexion of cause and effect was by no means the only idea by which the understanding thinks the connexion of this **a priori**, but rather that metaphysics consists altogether of such connexions. I sought to ascertain their number, and when I had satisfactorily succeeded in this by starting from a single principle, I proceeded to the deduction of these concepts, which I was certain were not deduced from experience, as Hume had apprehended, but sprang from the pure understanding." (Stroll and Popkins, 1972, p. 273)

An observer can acquire an improved understanding of sense experiences by seeking regularities and patterns among observations. Necessarily true knowledge — the raw material of science — emerges from this

Table 4.1. Historical Progression of Metaphysical Arguments from Descartes to Kant.

Descartes (1596–1650)	• Proponent of God, mind, and matter (dualist) • God: creator and preserver of mind and matter • Truth of all knowledge guaranteed by God • Dualist: mind and matter are distinct • Mechanistic (deterministic) view of nature
Hobbes (1588–1679)	• Proponent of God and matter (monist) • Mental processes are subset of matter • Deterministic view
Spinoza (1632–1677)	• Pantheistic view: all things are different forms of one substance • One substance: God
Leibniz (1646–1716)	• Matter consists of monads • Monads do not interact; they move in pre-established harmony • God orchestrates the behavior of monads • God is all-wise, all-good, all-knowing: therefore the world is the best of all possible worlds
Hume (1711–1776)	• Empiricist: relies on sense experience • Causal relations are empirical, not immutable
Kant (1724–1804)	• Scientific knowledge depends on the observed events and the observer • Metaphysical knowledge cannot be discovered by reason alone

process. Kant concluded that scientific knowledge depends on both the observer and the events observed.

Kant said we could not analyze substance relevant to metaphysics without making it subject to experimentation. At that point, our study of substance becomes empirical rather than metaphysical. The progression of metaphysical arguments discussed above is summarized in Table 4.1.

4.3 Classical Electrodynamics

Two fundamental forces were recognized before the 20th century: the gravitational force and the electromagnetic force. They were sufficient for describing the observed motion of objects. One fundamental force, the gravitational force, or gravity, affected the motion of any object with mass such as a ball rolling down an inclined plane or the motion of an astronomical body. The other fundamental force, the electromagnetic force, or electromagnetism, is associated with objects that have an electric charge. The electromagnetic force binds negatively charged electrons to positively charged atomic nuclei. No other forces were needed for pre-20th-century science, but experimental anomalies suggested that we need to learn more about the electromagnetic interaction to help us understand the need to modify Newtonian mechanics.

The first magnetic compasses were pieces of lodestone suspended so the lodestone piece could change its orientation in space. Lodestone is a naturally magnetized piece of the mineral magnetite. It seems to attract or repel other materials without touching them. For example, lodestone can attract iron filings, while two pieces of suspended lodestone can attract or repel each other depending on their orientation. These are examples of magnetic phenomena.

The presence of electric charge on a material is another phenomenon that appears to affect other materials without touching them. For example, it is possible to pick up small pieces of paper with a hard rubber rod in a dry climate if we rub the rod with fur and place the rod near small bits of paper. Another example of an electric phenomenon can occur when we walk across carpets during a cold, dry spell and touch a metal doorknob. The accumulation of electric charge on our bodies by walking on the carpet can result in an electric shock.

Electric and magnetic phenomena are special cases of a more general class of phenomena called electromagnetic phenomena. Englishman Michael Faraday (1791–1867) was one of the first scientists to conduct an extensive experimental study of electromagnetic phenomena. In addition to experimental contributions, Faraday made theoretical contributions to the study of electromagnetic phenomena. He introduced the concept of a field. A field has a value at each point in space and time. A magnetic field can be displayed by placing a magnet under a thin piece of paper or glass plate. The magnetic field appears when iron filings are sprinkled on the paper or plate. Another idea introduced by Faraday was the idea of force fields:

Q4.4. "One of Faraday's greatest intellectual innovations was the idea of force fields.... in the centuries between Newton and Faraday one of the great mysteries of physics was that its laws seemed to indicate that

> forces act across the empty space that separates interacting objects. Faraday didn't like that. He believed that to move an object, something has to come in contact with it. And so he imagined the space between electric charges and magnets as being filled with invisible tubes that physically do the pushing and pulling. Faraday called those tubes a force field." (Hawking and Mlodinow, 2010, p. 89)

French scholar Pierre-Simon Laplace (1749–1827) applied the idea of field to derive Newtonian gravity from a gravitational potential field. Laplace rewrote "Newton's theory of gravity so that the force was carried by a gravitational potential field" (Carroll, 2019, p. 248).

Faraday's experiments provided many observations that other scientists could use to construct a mathematical theory of electromagnetism. Scottish mathematician and physicist James Clerk Maxwell (1831–1879) devised the most successful mathematical formulation of electromagnetism. His set of equations described observations of electromagnetic phenomena provided by other researchers, including Faraday. Maxwell combined the mathematical description of electricity, magnetism, and light into a single set of equations. He showed that electric and magnetic fields are components of an electromagnetic field. Electric and magnetic forces were manifestations of a single electromagnetic force. In addition, Maxwell's equations showed that light was an electromagnetic wave, which is discussed in more detail next.

4.4 Wave Motion and the Electromagnetic Spectrum

Maxwell's equations can be used to show that light is a wave-like phenomenon. Figure 4.1 shows a single wave. The length of the wave from one point on the wave to an equivalent point is called the wavelength. The number of waves passing a particular point, say point B in Figure 4.1, in a specified time interval is the frequency of the wave.

A wave can be illustrated by attaching a rope to a screw eye in the wall. Hold the rope tight and move it up and then back to its original position. A pulse like that shown in Figure 4.2(A) should proceed toward the wall. The pulse is half of a wave and has one half a wavelength. When the pulse strikes the screw eye in the wall, it will be reflected as shown in Figure 4.2(B). To make a whole wave, move the rope up, back to its

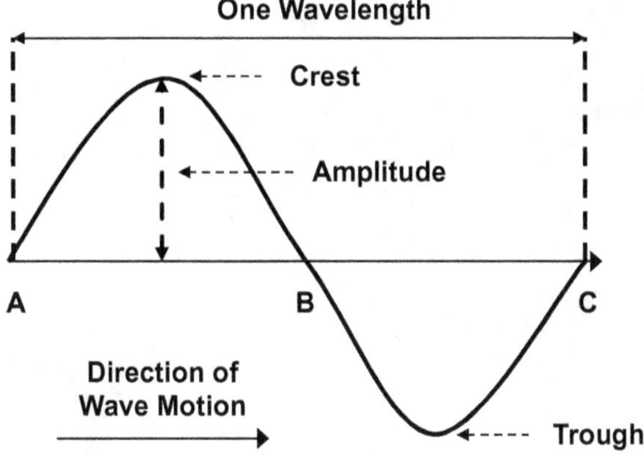

Figure 4.1. Sketch of a Wave.

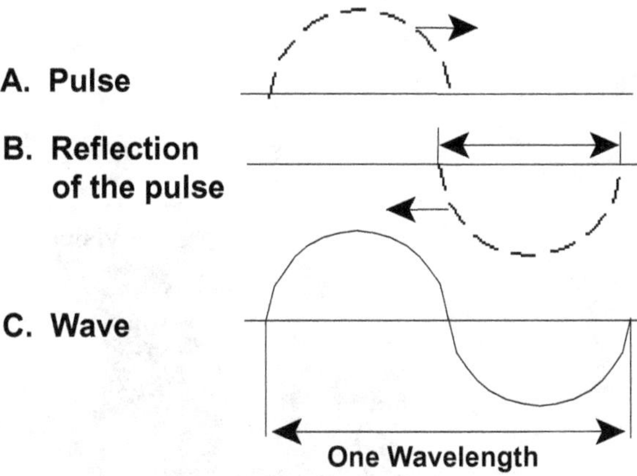

Figure 4.2. Sketch of a Wave Pulse.

original position, down, and then up to its original position. All of these motions should be made smoothly and continuously. The resulting pulses should look something like the complete wave in Figure 4.2(C). We can make many of these waves by moving the rope up and down rhythmically. The ensuing series of waves is called a wave train. The mathematical equation for describing the motion of a wave has the same form as the equation for describing the motion of light. Thus, light is often thought of as wave motion.

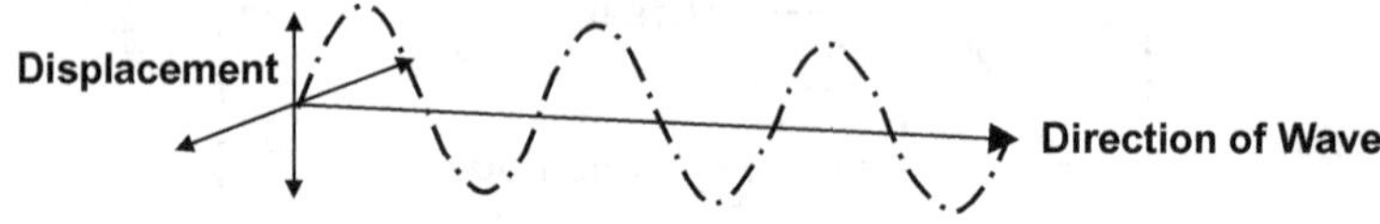

B. Light Wave

Figure 4.3. Sketch of a Sound Wave and a Light Wave.

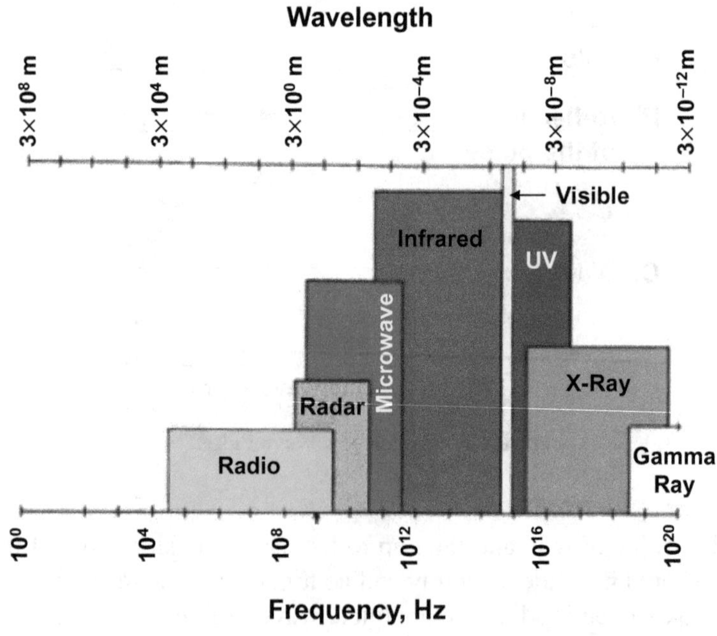

Figure 4.4. Electromagnetic Spectrum (1 Hz = 1 Cycle/Second).

Figure 4.3 compares a sound wave with a light wave. Sound is the wave-like transmission of vibrational energy. The displacement of the vibration is illustrated in Figure 4.3(A). Light, on the other hand, is the propagation of electric and magnetic fields, as shown in Figure 4.3(B). The electric field and the magnetic field oscillate in perpendicular directions relative to one another, and in a plane that is perpendicular to the direction of propagation of the resulting electromagnetic wave.

Different colors of light have different wavelengths and different frequencies. The wavelengths of light we see, visible light, represent only a very narrow band of wavelengths of the more general phenomenon known as electromagnetic radiation. The electromagnetic spectrum illustrated in Figure 4.4 shows several bands of electromagnetic radiation. For example, radio waves have relatively long wavelengths ranging from a fraction of a meter to over tens of kilometers. Visible light wavelengths are a thousand to a trillion times shorter than the wavelengths of radio waves. The wavelength of X-rays is one hundred to one thousand times smaller than the wavelength of visible light. Gamma rays, products of nuclear detonations, have wavelengths smaller than X-rays.

James Clerk Maxwell predicted that electromagnetic waves propagate at the speed of light. The upper axis of the electromagnetic spectrum in Figure 4.4 shows the wavelength of the light wave, and the lower axis shows the frequency of the light wave. The product of frequency and wavelength gives the speed of light in vacuum, which is 3×10^8 m/s. Maxwell's prediction of a large, finite light speed was confirmed by experiment.

# # #

Today, Maxwell's equations of electromagnetism are considered comparable in importance to Newton's laws of mechanics. Together, Newton's laws and Maxwell's equations form the theoretical basis of classical electrodynamics: the study of the motion of electrically charged objects in gravitational and electromagnetic fields. Classical electrodynamics, and especially the speed of light, played a fundamental role in the development of relativity.

Endnote

1. Sources on classical physics include D'Abro (1951), Bernal (1972), Jackson (1999), Morin (2007), Teerikorpi *et al.* (2019), and Leinaas (2019).

Part 2

Einsteinian Time

Chapter 5

Einstein's Special Relativity

The success of Newtonian mechanics was the basis for a deterministic view of the universe until the beginning of the 20th century. Newtonian mechanics is an example of a nonrelativistic theory because it is applicable to speeds much less than the speed of light.

Extension of mechanics to relativistic speeds, that is, speeds close to the speed of light, had to await experimental and theoretical advances in the late 19th and early 20th centuries. The historical context of relativity is discussed here. It includes Galileo's study of motion, incorporation of relativistic concepts into Newtonian mechanics, synchronizing travel by rail, and development of Einstein's theory of special relativity.[1]

5.1 Galileo's Ship

Galileo's understanding of time and interest in experiments led to advances in the understanding of ideas associated with motion. Motion can be discussed in terms of velocity. The velocity of a uniformly moving object has both magnitude and direction. The magnitude of velocity is often called speed. In many cases, it is sufficient to refer to speed when direction is of secondary importance. Other ideas associated with motion include inertia, frames of reference, and relativity. These ideas are discussed here in the context of Galileo's ship (Galilei, 1632, pp. 216–217), one of the first-thought experiments in relativity.

Galileo presented a discussion of the Ptolemaic geocentric view and the Copernican heliocentric view in *Dialogue Concerning the Two Chief World Systems: Ptolemaic and the Copernican*. The discussion was between three characters: Salviati, Simplicio, and Sagredo. Salviati was the proponent of the Copernican view supported by Galileo, Simplicio supported the Ptolemaic view adopted by the Roman Catholic Church, and Sagredo mediated the discussion.

Galileo chose the fictional character Salviati to present a thought experiment. He imagined what motion would be like to an observer enclosed in a ship traveling on a smooth sea with constant velocity. Salviati set the stage: "Shut yourself up with some friend in the main cabin below decks on some large ship, and have with you there some flies, butterflies, and other small flying animals. Have a large bowl of water with some fish in it; hang up a bottle that empties drop by drop into a narrow-mouth vessel beneath it" (Galilei, 1632, pp. 216–217). Today, we say that the observer in the interior of the ship is in a non-accelerating, or inertial, frame of reference.

Salviati then considered what the observer would see if the ship was at rest on the smooth sea: "With the ship standing still, observe carefully how the little animals fly with equal speed to all sides of the cabin. The fish swim indifferently in all directions; the drops fall into the vessel beneath; and, in throwing something to your friend, you need to throw it no more strongly in one direction than another, the differences being equal; jumping with your feet together, you pass equal spaces in every direction" (Galilei, 1632, p. 217).

Salviati changed the conditions of the thought experiment by allowing the ship to move with uniform speed relative to the sea. He described what the observer would see when the ship is moving: "When you have observed all these things carefully (though there is no doubt that when the ship is standing still everything must happen in this way), have the ship proceed with any speed you like, so long as the motion is uniform and not fluctuating this way and that. You will discover not the least change in all the effects named, nor could you tell from any of them whether the ship was moving or standing still" (Galilei, 1632, p. 217). The observer in the interior cabin of the ship was not able to observe the inertial motion of the ship on a smooth sea.

Salviati argued that the Copernican view of Earth moving around the Sun does not conflict with our everyday experience. For example, a ball falling from a tower falls vertically to the base of the tower; the ball does not follow a parabolic trajectory as Earth carries the tower away from the falling ball. After discussing similar observations of droplets falling in the cabin and fish swimming in a bowl, Salviati concluded that "the ship's motion is common to all things contained in it, and to the air also. That is why I said you should be below decks, for if this took place above in the open air, which would not follow the course of the ship, more or less noticeable differences would be seen in some of the effects noted" (Galilei, 1632, p. 218). Galileo, through Salviati, recognized that the observer standing on the ship's open deck is aware of the ship's motion relative to the smooth sea. Consequently, the observer in the interior cabin and an observer on the ship's open deck have different points of view or frames of reference.

5.2 Galilean Relativity

Galileo realized that laws of motion are the same in all inertial frames of reference, that is, frames of reference moving with constant velocity with respect to one another. It is worth noting that the motion of a ship in Galileo's thought experiment was an approximation of an inertial frame. In principle, a ship moving at a constant speed on Earth's spherical surface is accelerating because the ship is gradually changing direction. Inertial frames of reference do not accelerate.

Dutch scientist Christiaan Huygens (1629–1695) introduced the first clear statements of the principle of inertia and the relativity principle as Hypothesis I and Hypothesis III of his paper *On the Motion of Bodies Resulting from Impact* (Huygens, 1656). Hypothesis I stated the principle of inertia: "Any body once moved continues to move, if nothing prevents it, at the same constant speed and along a straight line" (Huygens, 1656).

Hypothesis III stated the relativity principle that inertial frames are equivalent: "The motion of bodies and their equal and unequal speeds are to be understood respectively, in relation to other bodies which are considered as at rest, even though perhaps both the former and the latter are involved in another common motion. And accordingly, when two bodies collide with each other, even if both together are further subject to another uniform motion, they will move each other with respect to a body that is carried by the same common motion no differently than if this motion extraneous to all were absent."

Huygens illustrated Hypothesis III using the example of an impact that takes place on a ship moving with constant speed. He asserted that the impact of colliding bodies on a moving ship was equivalent to the same impact taking place on the ship at rest. This illustration improved Galileo's ship example by expressing the illustration in the more precise language of Hypotheses I and III. The principle of Galilean relativity says that "mechanical experiments will have the same results in a system in uniform motion that they have in a system at rest" (DiSalle, 2020, Section 1.1). It is attributed to Galileo because he was the first to express the equivalence of inertial frames.

5.3 Newtonian Relativity

According to Newton, time and space are independent of each other and are permanent. A belief in the permanence of time and space implies the existence of points in space and instants of time that are independent of objects or observers confined within the space and time domains. From this perspective, an observer moving at a very great speed should measure the passage of time at the same rate as a motionless observer. For objects moving much slower than the speed of light, Newtonian space and time provided a satisfactory system of coordinates for specifying the location of an object in three-dimensional space at a given instant of time. The coordinate system used by an observer is a reference frame.

The observer on the deck of Galileo's ship does not have the same reference frame as an observer in the cabin below deck. The two reference frames are inertial frames of reference if they are moving with constant velocity relative to each other. Suppose another observer onshore is watching Galileo's ship go by. If the observer on the deck of the ship and the onshore observer are separating at a constant speed and in a straight line, they occupy inertial frames of reference. The frames of reference can be related by a transformation that allows us to transform between the coordinate system of the onshore observer and the coordinate system of the observer on the ship's deck. This transformation between the coordinates of the two reference frames depends only on the constant relative motion of the observers and is known as the Galilean transformation.

Invariance

A physical theory consists of a set of rules that relates mathematical quantities and operations to physical quantities and observations. Relationships between observations are typically embodied in a set of rules that is expressed as a mathematical formalism. The principle of relativity states that relationships between variables should remain invariant, or unchanged, with respect to transformations from one frame of reference to another. Invariance is demonstrated by showing that the equations of a physical theory retain their form when the coordinate system of one observer is transformed into the coordinate system of another observer.

Galilean invariance states that Newton's laws of motion should be the same in all inertial frames of reference. The Galilean principle of relativity says that the laws of motion are invariant with respect to Galilean invariance, that is, the laws of motion will be the same in a system at rest as they are in a system moving at constant velocity relative to the system at rest.

Another example of invariance is Lorentz invariance. This invariance principle was proposed as an explanation of the behavior of the speed of light in vacuum. Experimentalists discovered that the speed of light in vacuum does not depend on the motion of an inertial reference frame. An observer at rest will measure the same speed of light in vacuum as an observer that is measuring the speed of light in vacuum from an inertial reference frame that is moving at a speed that is much less than the speed of light. The constancy of the speed of light to observers in different inertial frames of reference was an anomaly of late 19th-century physics.

Maxwell's equations

James Clerk Maxwell (1831–1879) developed equations of electromagnetism that could be used to derive a wave equation for a wave that moved at the speed of light. Maxwell calculated the speed of light and interpreted the light wave as a wave moving through a medium called the luminiferous ether. The idea of a medium for light was based on analogy with sound waves in air and waves in water. Maxwell recognized that light from astronomical bodies must pass through the luminiferous ether on its way to Earth. The motion of Earth around the Sun means that it should be possible to detect different speeds of light at different times of the year as Earth orbits the Sun. Hawking and Mlodinow (2010, pp. 93–95) reported that Maxwell presented his idea to the editor of the *Proceedings of the Royal Society*, who chose not to publish Maxwell's idea because he did not think it would work. "In 1879, shortly before he died at age 48 of painful stomach cancer, Maxwell sent a letter on the subject to a friend. The letter was published posthumously in the journal *Nature*" (Hawking and Mlodinow, 2010, p. 95).

The Michelson–Morley experiment

Prussian-born American physicist Albert Michelson (1852–1931) read Maxwell's letter in *Nature* and was inspired to design and conduct an experiment in 1887 with American physicist Edward Morley (1838–1923). They measured the speed of light in vacuum at two different times of the year as illustrated in Figure 5.1. The timing of the measurements was designed to detect seasonal differences in the speed of light as Earth moved through the luminiferous ether.

Michelson and Morley did not detect the luminiferous ether. Instead, they observed that the speed of light did not change from season to season. Irish physicist George F. FitzGerald (1851–1901) was inspired to write a short letter to the editor of *Science* after reading about the Michelson–Morley experiment. His letter proposed an explanation of the unexpected result. He hypothesized that "the length of material bodies changes, according as they are moving through the ether or across it, by an amount depending on the square of the ratio of their velocity to that of light" (FitzGerald, 1889, p. 390). The object moving through the ether — an invisible fluid filling all space — would be compressed, or contracted, by the resistance of the ether to the motion of the object. Albert Einstein

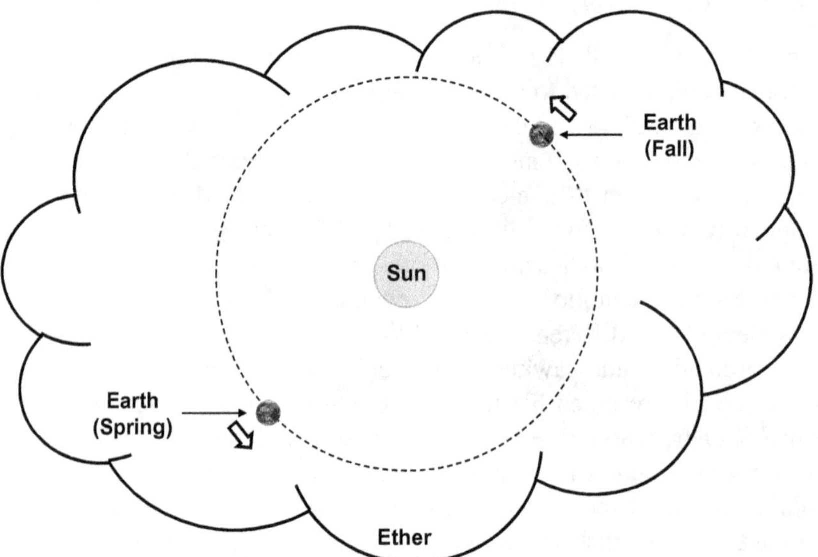

Figure 5.1. Earth's Motion through the Luminiferous Ether.

provided a different explanation for length contraction, which is discussed in Chapter 6 in the context of the pole-in-the-barn paradox.

The Michelson–Morley experiment is an example of an experiment with a negative result: the expected result was not observed, but the observed result was significant nonetheless. The negative result was eventually interpreted as evidence that the luminiferous ether did not exist.

5.4 Synchronizing Travel

Galileo's ship is an example of how travel at sea inspired people to think about motion in space and time. American physicist Adam Frank pointed out that travel by land and sea motivated improvements in the determination of time and location (Frank, 2011). Ships had circumnavigated the globe by the 17th century, and the problem of identifying your location at sea was a practical problem. Its solution required an ability to measure latitude and longitude.

Lines of constant latitude, or parallels, run from east to west around the world. Parallels form circles that are parallel to the equator. They specify the north–south position of a point on the Earth's surface. The latitude of a parallel at the equator is the angle 0° and increases to angle 90° at the poles. Latitude is measured by observing the Sun's highest position in the sky.

Longitude specifies the east–west position of a point on the Earth's surface. It is measured relative to a meridian, which is an arbitrary line that circles the globe and runs through both poles. An international agreement identified the meridian that runs through Greenwich, England as the prime meridian, that is, the meridian at longitude 0°. Longitude in the Eastern Hemisphere is measured in degrees east of the prime meridian, and longitude in the Western Hemisphere is measured in degrees west of the prime meridian.

A degree of longitude is approximately 111 km (69 mi) at its widest point near the equator.

A ship's longitude was measured by converting time measurements to locations on the Earth. Local time when the Moon reached its highest point in the sky was recorded and compared to a book that listed times when the moon reached its highest point at the prime meridian in Greenwich, England. Local time on the ship was obtained by reading a clock calibrated to the local position of the Sun. A comparison of the Greenwich time with the local time transformed the local time into a local position relative to the prime meridian. The conversion of a local time to a local position relied on the observation of a simultaneous event — the time when the moon reached its highest point in the sky — on two different meridians.

Travel by rail

Travel by land was revolutionized with the invention of travel by rail. By the end of the American Civil War in 1865, railroads "were in the middle of their assault on the continents. In that same year, telegraph cables threading electrical impulses into instantaneous communication were just beginning to bind far-flung cities to each other" (Frank, 2011, p. 119). A new practical problem emerged: how can we ensure that train schedules are met at different locations in a railroad network?

Large cities established their own time standards in the latter half of the 19th century. "The clocks of each city were set according to a regional time standard often provided, remarkably, by astronomers working at the local university observatory" (Frank, 2011, p. 120). Regional time standards typically defined local noon as the time when the Sun reached its highest point in the sky. Different regional time standards complicated travel by rail when a traveler passed through regions with different time standards. A meeting called the General Time Convention was convened in Chicago on October 11, 1883, to discuss the issue.

Reformers at the General Time Convention wanted to move time standards away from their dependence on the local apex of the Sun. Others at the Convention preferred to retain the current time standards. The reformers offered a "compromise between the economic needs of a nation stitched together by train lines and the natural local rhythms of day and night. They divided the nation up into time zones, each 15° wide" (Frank, 2011, p. 122). If we divide the circumference of a circle (360°) by 15° per time zone, we obtain 24 time zones around the world corresponding to 24 hours in a day. The major railroads supported the plan, which was subsequently nationalized by legislation.

Telegraph lines were developed concurrently with the development of railroad systems. The telegraph made it possible for people separated by a great distance to simultaneously exchange electromagnetic signals. The simultaneous exchange of information and the existence of time zones facilitated synchronization of time and the preparation of precise maps. Mapmaking was an important business in the latter half of the 19th century as several nations engaged in empire-building.

"At the very moment when the world was desperate to construct networks of electromagnetic simultaneity, the young Albert Einstein was hard at work for the Swiss Patent Office. His day job was to evaluate designs for electromechanical time coordination devices. [His night job was to] weave simultaneity into a theory of space, time, matter and energy" (Frank, 2011, p. 125).

5.5 Einsteinian Relativity

Albert Einstein's (1879–1955) interest in time and simultaneity helped him develop an explanation of the constancy of the speed of light in a 1905 paper entitled *On the Electrodynamics of Moving Bodies* (Einstein, 1905a). Einstein was aware of the luminiferous ether, but considered it superfluous: "The introduction of a 'luminiferous ether' will prove to be superfluous inasmuch as the view here … will not require an 'absolutely stationary space' provided with special properties" (Einstein, 1905a, p. 38). American physicist Richard Muller pointed out that there is a question about how much Einstein was influenced by the 1887 Michelson–Morley experiment. Muller said that "it appears to many that his theory of relativity was based primarily on the properties of the Maxwell theory of electromagnetism, and the properties of that theory deduced by Lorentz" (Muller, 2016, p. 40).

Before Einstein's 1905 paper, physicists worked with invariance and transformation requirements associated with Galilean relativity. Einstein proposed a new set of transformation rules collectively known as the Lorentz transformation after Dutch physicist Henrik Anton Lorentz (1853–1928). The Lorentz transformation kept the form of Maxwell's equations unchanged, or invariant, with respect to an observer in an inertial reference frame. This required the replacement of Newton's independent 'absolute space' and 'absolute time' coordinates with space and time coordinates that were mutually dependent. The Lorentz transformation adopted by Einstein ensured the constancy of the speed of light when one inertial reference frame was transformed into another.

If we examine the invariance properties of Newtonian mechanics and Maxwell's equations of electromagnetism, we find that Newton's laws of motion are invariant with respect to Galilean invariance, but are not

Lorentz invariant. On the other hand, Maxwell's equations are Lorentz invariant, but they are not Galilean invariant. There was a clear contradiction between the two most successful physical theories known prior to the 20th century.

The contradiction resides in Newton's conception of space and time. Newton thought space and time were absolute and independent quantities, but the failure of Maxwell's equations to be invariant with respect to a Galilean transformation implied that Newtonian mechanics could be flawed at speeds approaching the speed of light. Einstein decided to reject the concept of 'absolute time':

> **Q5.1**. "Every reference-body (coordinate system) has its own particular time; unless we are told the reference-body to which the statement of time refers, there is no meaning in a statement of the time of an event." (Einstein, 1924, p. 32)

The term 'event' refers to a point in space at a specified time, where space is measured by rulers and time is measured by clocks. The elapsed time, or duration, between two events has meaning only if we specify the reference frame in which the duration was measured.

Einstein recognized that the concept of time needed to be reconsidered.

> **Q5.2**. "We have to take into account that all our judgments in which time plays a part are always judgments of simultaneous events." (Einstein, 1905a, p. 39)

American historian of science James Gleick noted that Einstein was not the first to question Newton's absolute time. According to Gleick,

> **Q5.3**. "The philosopher and physicist Ernst Mach, a forebear of relativity, objected to absolute time in 1883: 'It is utterly beyond our power to measure the changes of things by time… Time is an abstraction at which we arrive by means of the changes of things.' Einstein quoted that approvingly when he wrote Mach's obituary in 1916, but he himself could not go so far in expunging the convenient abstraction. Time remained an essential property of his universe." (Gleick, 2016, p. 92, footnote 2)

Einstein's revision of the concept of time depended on synchronizing clocks at three points A, B, and C:

> **Q5.4**. "1. If the clock at B synchronizes with the clock at A, the clock at A synchronizes with the clock at B.
>
> 2. If the clock at A synchronizes with the clock at B and also with the clock at C, the clocks at B and C also synchronize with each other." (Einstein, 1905a, p. 40)

Einstein used the concept of synchronicity to define the time of an event as

> **Q5.5**. "that which is given simultaneously with the event by a stationary clock located at the place of the event, this clock being synchronous and indeed synchronous for all time determinations, with a specified stationary clock." (Einstein, 1905a, p. 40)

The simultaneity of two events that are separated in space depends on the motion of the observer:

> **Q5.6**. "We cannot attach an ***absolute*** signification to the concept of simultaneity, but that two events which, viewed from a system of coordinates, are simultaneous, can no longer be looked upon as simultaneous events when envisaged from a system which is in motion relatively to the system." (Einstein, 1905a, p. 42)

The dependence of simultaneity on the motion of the observer becomes significant for speeds comparable to the speed of light.

Given his new view of space, time, and simultaneity, Einstein was able to resolve the constancy of the velocity of light anomaly by specifying two postulates: a principle of relativity and a principle of the constancy of the velocity of light. The first postulate is the principle of relativity:

> **Q5.7**. "The laws by which the states of physical systems undergo change are not affected whether these changes of state be referred to the one or the other of two systems of coordinates in uniform translatory motion." (Einstein, 1905a, p. 41)

The relationships between physical quantities should not depend on the relative motion of observers.

The second postulate is the principle of the constancy of the velocity of light:

> **Q5.8**. "Any ray of light moves in the 'stationary' system of coordinates with the determined velocity *c*, whether the ray be emitted by a stationary or by a moving body." (Einstein, 1905a, p. 41)

Rather than considering the constancy of the velocity of light an anomaly, Einstein elevated the constancy of the velocity of light to a postulate.

The speed of light is 299,792,458 m/s, which is approximately 300,000 km/s or 186,000 mi/s. Light speed is large but finite. If it is true that nothing can travel faster than the speed of light, then two events can interact with each other only if the events are close enough that a signal emitted by one event can arrive at the other event by traveling at a speed less than or equal to the speed of light. The interaction between events is considered a causal relationship.

Addition of relativistic velocities

According to Einstein's special theory of relativity, the speed of light is the maximum speed that can be attained by an object. In this case, the addition of two velocities using Newtonian mechanics cannot be correct. We can demonstrate the addition of relativistic velocities by imagining a motionless observer watching two spaceships approach each other. To be specific, suppose Bob is traveling in a spaceship at half the speed of light relative to Max, a stationary observer on Earth, as sketched in Figure 5.2. We could just as easily think of Bob as the stationary observer and Max as the observer moving at half the speed of light. Either point of view is acceptable in principle. For our example, it is more convenient to think of Max as stationary.

We observe Alice approaching Bob in another spaceship. Alice's speed relative to Max is two-thirds the speed of light. According to Newton's view of space and time, Max should observe Bob approaching Alice at a speed in excess of the speed of light. If Max actually measured the speed with which Bob is approaching Alice as observed from Max's stationary reference frame, he would find the speed of approach of the two

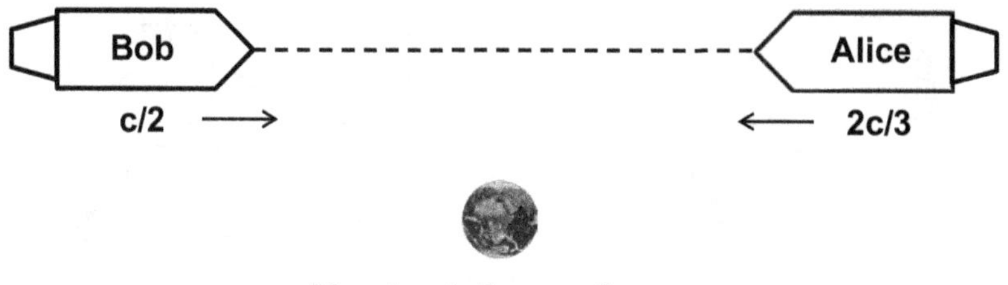

Figure 5.2. Sum of Relativistic Velocities.

spaceships to be less than the speed of light. In other words, Max's measurement does not agree with Newton's theory. What is wrong with the classical sum of relativistic speeds?

The addition of relativistic velocities must account for the Lorentz transformation. When the Lorentz transformation is considered, the sum of two velocities does not exceed the speed of light.

Relativistic mass

The theory presented by Einstein in his 1905 paper is known today as the 'special theory of relativity'. It is concerned with reference frames moving uniformly with respect to other reference frames. These uniformly moving reference frames are known as non-accelerating, or inertial, reference frames. Einstein showed that the special theory of relativity implied that "the mass of a body is a measure of its energy content" (Einstein, 1905b, p. 71). According to Einstein,

> **Q5.9**. "the most important result of a general character to which the special theory of relativity has led is concerned with the conception of mass. Before the advent of relativity, physics recognized two conservation laws of fundamental importance, namely, the law of the conservation of energy and the law of the conservation of mass... By means of the theory of relativity [the laws of conservation of mass and energy] have been united into one law." (Einstein, 1924, p. 45)

The unified law referred to by Einstein is now known as the law of conservation of mass–energy. The equivalence of mass and energy is the theoretical basis of nuclear power and nuclear weapons.

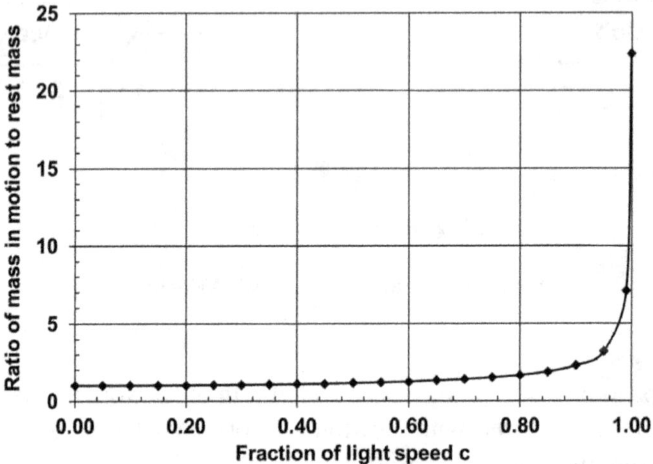

Figure 5.3. Relativistic Mass from Rest to 0.999*c*.

Einstein's special theory of relativity showed that the mass of a moving object increases as the speed of the object increases. Figure 5.3 shows that the ratio of the mass of a moving object divided by its rest mass, or mass at rest relative to its frame of reference, increases as the speed of the mass approaches the speed of light *c*. The last plotted point on the right-hand side of Figure 5.3 is at a speed of 0.999*c*. In theory, the mass of the object in motion becomes infinite at the speed of light. This implies that no massive object can move at a speed greater than the speed of light, which seems to establish the speed of light as the upper limit on the maximum possible speed of any object.

# # #

The special theory of relativity applies only to non-accelerating reference frames. Einstein did not present a theory of relativity for reference frames which were accelerating with respect to one another until 1916. The 1916 theory is called the 'general theory of relativity' and applies to accelerating bodies, such as a massive object moving in a gravitational field. As Einstein tells us, the general theory applies to a broad range of reference systems: "The laws of physics must be of such a nature that they apply to systems of reference in any kind of motion" (Einstein, 1916,

p. 113). Before discussing the general theory of relativity, it is important to first review important paradoxes associated with the special theory of relativity.

Endnote

1. Sources on special relativity include Sommerfeld (1923), Pais (1982), Stachel (1998), Hawking and Mlodinow (2010), Frank (2011), Gleick (2016), Muller (2016), and Teerikorpi *et al.* (2019).

Chapter 6

Paradoxes of Special Relativity

Special relativity seems to contain many contradictions and paradoxes[1] that help illuminate the meaning of the theory. For example, the pole-in-the-barn paradox is based on Lorentz contraction, the twin paradox depends on time dilation, and the grandfather paradox illustrates time travel issues. They are discussed here following an introduction to spacetime diagrams.

6.1 Spacetime Diagrams

Hermann Minkowski (1864–1909), Albert Einstein's former math teacher, reviewed Einstein's papers and realized that he could provide a geometric interpretation of Einstein's treatment of space and time. Minkowski was the first to show that Einstein's special theory of relativity unified space and time into a spacetime continuum which included one time and three space coordinates:

> **Q6.1**. "Henceforth space by itself, and time by itself, are doomed to fade away into mere shadows, and only a kind of union of the two will preserve an independent reality." (Minkowski, 1908, p. 75)

The special theory of relativity replaced Isaac Newton's view of absolute and independent space and time with a four-dimensional spacetime continuum.

Four numbers are needed to specify a point in spacetime: three space coordinates and one time coordinate. The set of four numbers defines a world point or an event in spacetime. The set of all world points that depict the trajectory of an object in spacetime is called the worldline. We cannot draw a four-dimensional diagram, but we can draw simpler diagrams.

The spacetime diagram in 1 + 1 dimensions shown in Figure 6.1 has one space axis (x-axis) and one time axis (t-axis). Time is measured along the time axis from past in the lower half of the diagram to future in the upper half of the diagram. The present moment is the event at the origin of the axes. The axes define a coordinate system that can display the trajectory of an object in spacetime. An event is a point on the spacetime diagram. The path of an object in spacetime is traced by a sequence of events called the worldline.

Another useful diagram is the spacetime diagram in 2 + 1 dimensions shown in Figure 6.2. Two space axes (x-axis and y-axis) and a time axis (t-axis) define a reference frame, or coordinate system, that can display the trajectory of events in spacetime. The trajectory of events in Figure 6.2 is the worldline that begins in the past light cone (lower cone) and ends in the future light cone (upper, inverted cone).

Figures 6.1 and 6.2 illustrate regions of spacelike, timelike, and lightlike separation. Suppose two events A and B are separating from each other at the speed of light. Both events are on the lightlike boundaries of the light cones. The inverted upper cone is the future light cone and the lower cone is the past light cone. The point where the tips of the cones in Figure 6.2 meet is the present. If two events A and B are separating at a

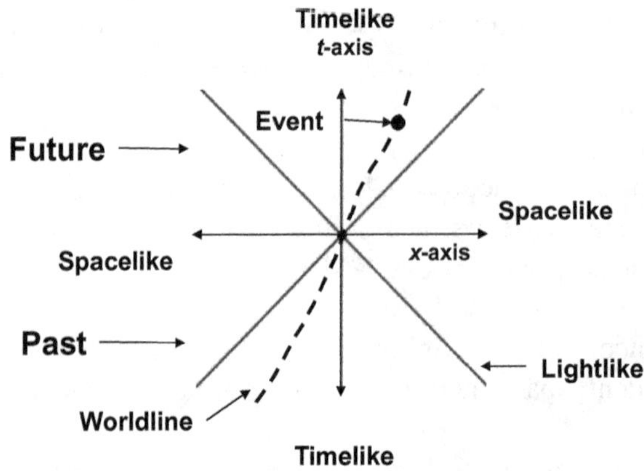

Figure 6.1. Spacetime Diagram in 1 + 1 Dimensions.

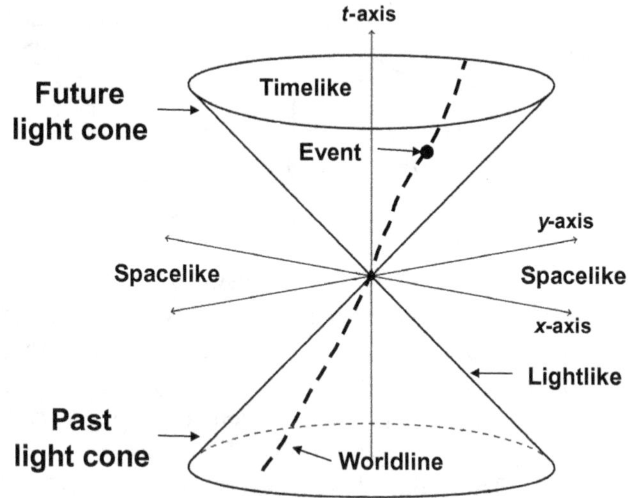

Figure 6.2. Spacetime Diagram in 2 + 1 Dimensions.

speed less than the speed of light and event A is in the timelike region, then event B is in the timelike region of event A. Suppose two other events C and D are separating at a speed greater than the speed of light and event C is in the timelike region, then event D is in the spacelike region of event C's light cone. Spacelike, timelike, and lightlike regions receive more attention in subsequent chapters.

The terms defined for light cones in Figure 6.2 are applicable to the triangles in the upper half and lower half of Figure 6.1. Light cones are represented by triangles in Figure 6.1 because the 1 + 1 spacetime diagram only has one space axis. The 1 + 1 spacetime diagram in Figure 6.1 is an example of a Feynman diagram, which is discussed in more detail later.

6.2 Pole-in-the-Barn Paradox

Two events are simultaneous if they occur at the same time. If two reference frames are in motion relative to each other, the timing of the two events depends on the distance separating the events. The separation Σ_1 between two events observed in a stationary reference frame F_1 is less than the separation Σ_2 between the events observed in reference frame F_2 moving at a constant velocity v relative to F_1.

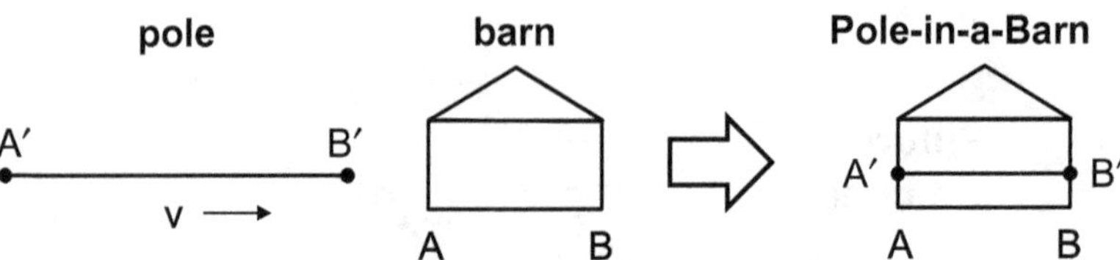

Figure 6.3. Pole-in-the-Barn Paradox.

The effect of length contraction is illustrated in Figure 6.3 for the pole-in-the-barn paradox. The left-hand side of the figure shows the pole approaching the stationary barn. We assume the barn is a stationary reference frame F_1 relative to reference frame F_2 moving with the pole at a constant velocity v relative to the barn. Reference frame F_2 is the proper frame of the pole, that is, the frame in which the observer is at rest with respect to the pole. The proper length of the pole from A′ to B′ is the length of the pole measured by an observer in F_2. In our example, the proper length of the pole from A′ to B′ is twice as long as the width of the barn from A to B. Is it possible to simultaneously fit both ends of the pole-in-the-barn, as shown on the right-hand side of the figure?

The Lorentz transformation shows that the length of the moving pole from A′ to B′ contracts when viewed from the reference frame of the barn as the velocity of the pole v increases. The length contraction is known as the Lorentz contraction or Lorentz–FitzGerald contraction, where Irish physicist George F. FitzGerald is recognized for suggesting that the length of an object compresses when moving through the ether (FitzGerald, 1889). If velocity v is large enough, the contraction of the pole length can allow the pole to fit inside the barn from the point of view of the stationary reference frame F_1. The word 'inside' means that both ends of the pole are in the barn at the same time as measured by an observer in F_1. To an observer in the proper frame of the pole, that is, riding with the pole, the front of the pole hits the barn wall at A before the back of the pole enters the barn. Observers in F_1 and F_2 do not agree that the ends of the pole are in the barn simultaneously. In this example, an observer in F_1 would see both ends of the pole simultaneously inside the barn when v is approximately 87% of the speed of light c.

6.3 The Twin Paradox

The pole-in-the-barn paradox is a consequence of the Lorentz contraction. The twin paradox is associated with time dilation, that is, the lengthening of time duration associated with the relative motion of two reference frames. The duration T_1 between two events observed in a stationary reference frame F_1 is longer than the duration T_2 between the events observed in reference frame F_2 moving at a constant velocity v relative to F_1.

There is considerable evidence for time dilation. For example, muon particles have a relatively short lifetime. The lifetime of a particle is the length of time a particle is stable before it transforms into other particles. Cosmic ray muons are created in the upper atmosphere of Earth when cosmic-ray protons collide with molecules in the atmosphere. The lifetime of a cosmic ray muon is too short to be stable long enough to reach muon detectors on the Earth's surface if the muon is moving too slowly. On the other hand, a cosmic ray muon moving at speeds close to the speed of light can be stable long enough to reach surface muon detectors. Figure 6.4 illustrates the distance traveled by a cosmic ray muon μ with proper lifetime Λ_0, the lifetime measured from a reference frame traveling with the muon, and the lifetime Λ that has been extended by time dilation.

The twin paradox is illustrated in Figure 6.5. We follow the journey of two identical twins Carol and Dawn. Carol is an astronaut who travels

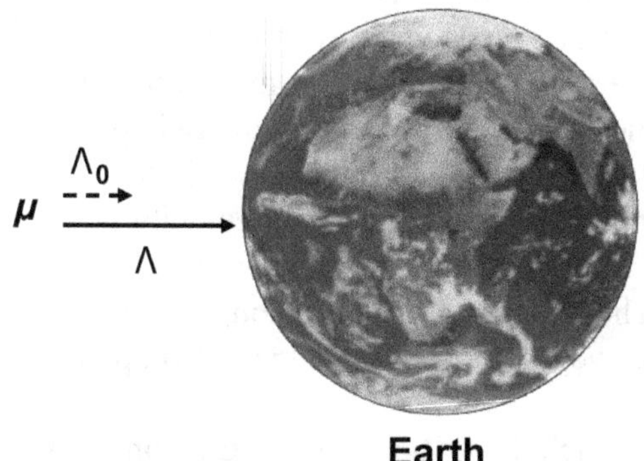

Figure 6.4. Lifetimes of Cosmic Ray Muons.

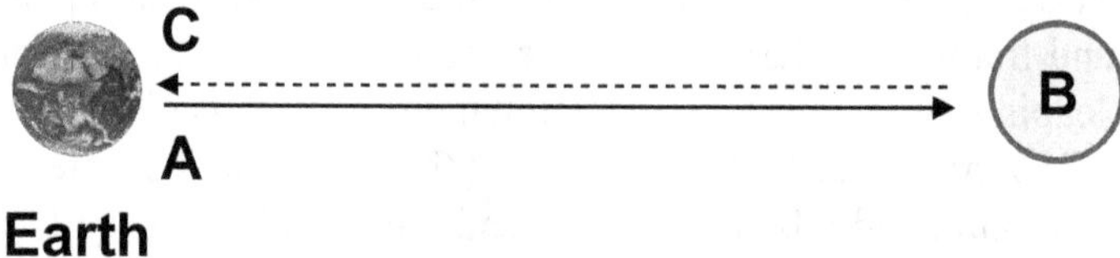

Figure 6.5. The Twin Paradox.

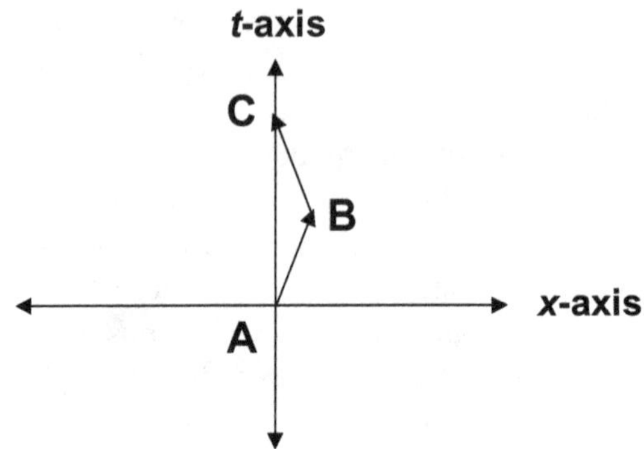

Figure 6.6. Spacetime Diagram for the Twin Paradox.

from A → B → C, while Dawn stays home as an observer in a stationary reference frame. The segment of the trip from A → B is outbound from Earth, and the segment from B → C is inbound to Earth. Point B is Carol's turnaround point from outbound to inbound. If Carol is traveling at a constant relativistic velocity v on each segment of the trip, which twin is older by the time Carol returns to Earth?

The spacetime diagram for the twin paradox is shown in Figure 6.6. Dawn follows the path from A → C while Carol follows the path from A → B → C. One approach to resolving the twin paradox is to argue that special relativity does not apply to this case. Special relativity is a theory of reference frames moving at a constant velocity relative to one another. The turnaround point B in the twin paradox is a change in the direction of Carol's speed. The change in the direction of velocity is an acceleration and says that the two reference frames in the twin paradox accelerate relative to each other.

Another approach is to apply time dilation to the twin paradox, including the turnaround. Carol turns around by stopping, reversing direction, and returning to Earth. Carol's proper frame is accelerated by the turnaround, while Dawn's proper frame does not accelerate.

Carol occupies two different proper frames: an outbound frame and an inbound frame. The duration of time between the two proper frames is called 'leaping time', or loss of simultaneity, by physicist Richard A. Muller (2016, pp. 54–55). Muller says that there is a leaping time between Carol's outbound and inbound frames, and there is also a loss of simultaneity between Carol and Dawn as a result of the acceleration during Carol's journey. Muller calculated that Carol would be younger than

Dawn, the twin who stayed on Earth, when Carol returned to Earth. Muller advised that special relativity should not be applied to accelerating frames, but if it is applied to a frame that accelerates, "then you have to take into account the leaping time of distant events" (Muller, 2016, p. 55).

6.4 The Grandfather Paradox

One of the objectives of defining a spacetime continuum is to treat space coordinates and time coordinates in the same way. We know that motion in space is possible: we can move forward and backward in space. Is it possible to move forward and backward in time? Is time travel possible? The grandfather paradox is an attempt to understand time travel in the context of special relativity.

Figure 6.7 illustrates key aspects of the grandfather paradox:

- Al has a son Ben
- Ben has a son Carl
- Carl travels back in time before Al has children
- Carl kills Al.

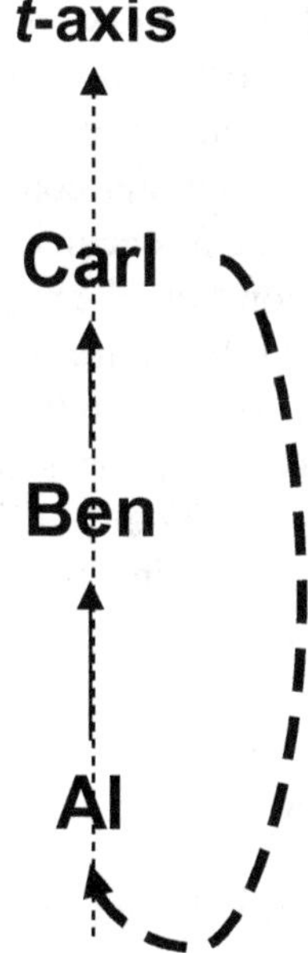

Figure 6.7. Time Travel and the Grandfather Paradox.

Some versions of the grandfather paradox say that the grandson (Carl) invented the time machine, so he could go back in time. Once the grandson killed the grandfather (Al), a new worldline began in which the father (Ben) was not born, the grandson (Carl) was not born, and Carl's time machine was not invented.

The grandfather paradox asks the question: can an effect precede its cause? Historian of science James Gleick concluded from his review of the time travel literature that time travel paradoxes "stem from retrocausality. Effects undo their causes" (Gleick, 2016, p. 235). In our version of the grandfather paradox, Carl (an effect) undid his cause (grandfather Al).

Physicist Richard A. Muller suggested one possible resolution of the grandfather paradox. According to Muller, "One possible answer is that you don't have free will, so even if you went back in time, you could not kill your grandfather. And the fact that you were eventually born shows you didn't do that" (Muller, 2016, p. 251).

Historian of science James Gleick said that *The Time Machine* by H.G. Wells "revealed a turning in the road, an alteration in the human relationship with time" (Gleick, 2016, p. 291). *The Time Machine* was published in 1895 (Wells, 1895), years before Einstein published the special theory of relativity in 1905 and Hermann Minkowski introduced spacetime as a geometric interpretation of space and time in 1908. Wells' fictional character the Time Traveller (sic) believed that "There is no difference between Time and any of the three dimensions of Space except that our consciousness moves along it" (Wells, 1895, p. 4). Gleick acknowledged a link between time and consciousness. He defined consciousness as "a dynamical system, occurring in time, evolving in time, able to absorb bits of information from the past and process them, and able as well to create anticipation for the future" (Gleick, 2016, p. 267). Gleick described Mental Time Travel as the "ability to mentally project oneself backward in time to relive past experiences and forward in time to relive possible future experiences." (*Ibid.*, p. 278) According to Gleick, "Everyone seems to agree that our imaginations liberate us in the time dimension" (*Ibid.*, p. 279).

6.5 The Meaning of 'Now'

Augustine of Hippo (354–430) believed that an eternal God is timeless. He recognized three coexisting presents:

> **Q6.2.** "The time present of things past is memory; The time present of things present is direct experience; the time present of things future is expectation." (Augustine, 398, Book XI, Chapter XX)

American physicist Richard A. Muller compared Augustine's view of three coexisting presents with a modern interpretation of physics that "describes the behavior of objects within time in spacetime diagrams that make no reference to the fact that time flows or that a now exists" (Muller, 2016, p. 18).

The time axis of the spacetime diagram has three distinct sections: past, present, and future. There are different views of the relationship between past, present, and future. Presentism is

> **Q6.3.** "the idea that only the present is real, that the past and the future are not — and that reality evolves from one present to another, successive one." (Rovelli, 2017, p. 106)

The present is the moment on the spacetime diagram between all past moments and all future moments. A line of successive present moments forms a worldline.

Another view of the time axis is eternalism, which is

> **Q6.4.** "the idea that flow and change are illusory: present, past, and future are all equally real and equally existent. Eternalism is the idea that the whole of spacetime … exists all together in its entirety without anything changing. Nothing really flows." (Rovelli, 2017, p. 108)

Einstein seemed to support the idea of eternalism in a March 1955 letter of condolence he wrote to the son and sister of his long-time friend and colleague Michele Besso. In the letter, Einstein wrote that

> **Q6.5.** "now he has departed from this strange world a little ahead of me. That means nothing. People like us who believe in physics, know that the distinction between past, present and future is only a stubbornly persistent illusion." (Rovelli, 2017, pp. 108–109)

The idea that the distinction between past, present, and future is illusory suggests that the passage from one moment of time to another is illusory. Instead, we should think of the history of the universe as a single block. Such a universe is called a 'block universe' and is an eternalist perspective. The block universe is discussed in Chapter 15.

The block universe does not account for the passage of time, the importance of the present moment *Now*, the apparent direction of time, or the difference between past and future. The only moment in which we can make a choice is the present moment *Now*. If we are only considering a choice, the decision is in our future, while the decision becomes part of our past once the decision is made (Muller, 2016).

# # #

Special relativity applies to non-accelerating reference frames. The need to understand gravity and the physics of accelerating reference frames led Einstein to develop a new theory: general relativity. General relativity is a theory of gravity that has replaced Newton's view of gravity, which is discussed next.

Endnote

1. Sources on paradoxes in special relativity include Wheeler (1990), Frank (2011), Muller (2016), and Rovelli (2017).

Chapter 7

Gravity and Action at a Distance

Our historical understanding of planetary motion is outlined in Table 7.1. It covers the period from Aristarchus of Samos to English scientist Isaac Newton. Our modern understanding of planetary motion emerged from the observations of Brahe and Copernicus, Johannes Kepler's three empirical laws of planetary motion, and the ability to derive Kepler's laws from the more generalized laws of Newtonian mechanics.

A few anomalies remained, such as motion through the luminiferous ether and details about the orbit of the planet Mercury. The anomalies were eventually explained using Albert Einstein's (1879–1955) general theory of relativity, but only after significant changes in our understanding of planetary motion. In this chapter, we describe Newton's understanding of gravity and planetary motion.[1] He introduced the concept of action at a distance, which made many of his contemporaries uncomfortable. We discuss the relative merits of action at a distance and fields. This discussion provides a primer for understanding Einstein's general theory of relativity.

7.1 Newtonian Gravity

Isaac Newton defined gravity in terms of centripetal force:

> **Q7.1**. "Definition V: A centripetal force is that by which bodies are drawn or impelled, or any way tend, towards a point as to a center." (Newton, 1687, p. 2)

Table 7.1. Historical Understanding of Planetary Motion.

Aristarchus of Samos (ca. 310–230 BCE)	Heliocentric hypothesis
Claudius Ptolemy (ca. 100–170)	Geocentric hypothesis
Nicolaus Copernicus (1473–1543)	Heliocentric hypothesis
Tycho Brahe (1546–1601)	Accurate observations; model with Earth at center, Sun orbiting Earth, other planets orbiting Sun
Johannes Kepler (1571–1630)	Laws of planetary motion derived from Brahe's measurements
Galileo Galilei (1564–1642)	Adopted heliocentric model; used telescope for observations
Isaac Newton (1643–1727)	Laws of motion; gravity and action at a distance

To Newton, gravity was the centripetal force that attracted bodies to the center of the earth. Magnetism was considered a centripetal force that attracted iron filings to a magnetic lodestone. Newton described the force responsible for planetary orbits as

> **Q7.2**. "that force, whatever it is, by which the planets are continually drawn aside from the rectilinear motions, which otherwise they would pursue, and made to revolve in curvilinear orbits." (Newton, 1687, p. 2)

Notice that Newton did not identify gravity as the force that kept planetary bodies in their orbits.

Each of these examples is an example of action at a distance, which is discussed in more detail below. Other examples of centripetal force discussed by Newton included a whirling stone, essentially a slingshot, and projectile motion. Newton described the slingshot as a stone, whirling about in a sling that is being turned by hand. As soon as the stone is released, it flies away. According to Newton,

> **Q7.3**. "that force which opposes itself to this endeavor, and by which the sling continually draws back the stone towards the hand, and retains it in its orbit, because it is directed to the hand as the center of the orbit,

> I call the centripetal force. And the same thing is to be understood of all bodies, revolved in any orbits." (Newton, 1687, pp. 2–3)

Newton viewed the slingshot as an illustration of orbital motion. He generalized the concept to include all bodies that

> **Q7.4**. "endeavor to recede from the centers of their orbits; and were it not for the opposition of a contrary force which restrains them to, and detains them in their orbits, which I therefore call centripetal, would fly off in right lines, with a uniform motion." (Newton, 1687, pp. 2–3)

Newton applied the ideas associated with orbital motion to projectile motion. He first recognized that the trajectory of a projectile would not bend towards the earth unless it was acted upon by the force of gravity:

> **Q7.5**. "A projectile, if it was not for the force of gravity, would not deviate towards the earth, but would go off from it in a right line, and that with a uniform motion, if the resistance of the air was taken away." (Newton, 1687, p. 3)

He then considered a leaden ball that was projected from the top of a mountain by the force of exploding gunpowder. The trajectory of the ball was initially in a direction parallel to the horizon, but the trajectory bent toward the horizon because of the force of gravity, even if we neglect air resistance. The trajectory depends on the initial velocity of the projectile:

> **Q7.6**. "By increasing the velocity, we may at pleasure increase the distance to which it might be projected, and diminish the curvature of the line which it might describe, till at last it should fall… or even might go quite round the whole earth before it falls." (Newton, 1687, p. 3)

Newton extended these ideas to trajectories with initial velocities that would propel the ball

> **Q7.7**. "so that it might never fall to the earth, but go forwards into the celestial spaces, and proceed in its motion *in infinitum*." (Newton, 1687, p. 3)

Newton recognized that a body that is propelled beyond the earth could go into orbit around the earth and even into orbit around the moon. Almost two centuries later, a similar idea was embraced by French author Jules Verne in his 1867 novel *From the Earth to the Moon*.

7.2 Action at a Distance

Isaac Newton's law of universal gravitation was widely considered the correct mathematical description of gravity by the scientific community at the beginning of the 20th century. In the Newtonian model, gravity was the result of an instantaneous and direct interaction between two masses. As an illustration, Figure 7.1(A) shows an object with mass m_1 being attracted by a gravitational force to the much more massive Earth with mass m_E. The gravitational force is an example of action at a distance. To Newton, action at a distance refers to changes in the motion of object A that occur even though object A has not been in mechanical contact with another object B, that is, the motion of object A is not produced by direct contact with another object.

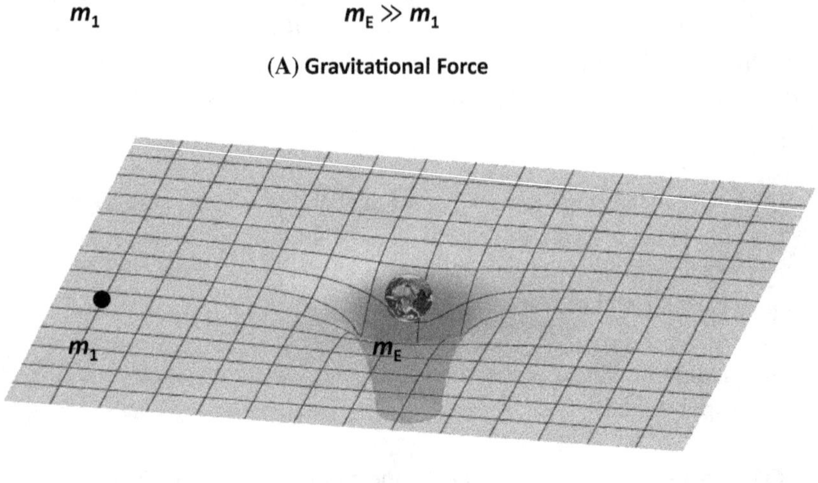

Figure 7.1. Representing Gravity as a Force and a Field.

Many of Newton's contemporaries did not believe that action at a distance was possible. Even Newton was skeptical. In a letter dated February 25, 1692–1693 to the Reverend Doctor Richard Bentley, Newton wrote

> **Q7.8**. "it is inconceivable that inanimate brute Matter should, without the Mediation of something else, which is not material, operate upon, and affect other matter without mutual Contact." (Newton, 1692–1693, p. 25)

Newton did not want to be associated with the idea of innate gravity. He explained that

> **Q7.9**. "this is one Reason why I desired you would not ascribe innate Gravity to me. That Gravity should be innate, inherent and essential to Matter, so that one Body may act upon another at a Distance thro' a Vacuum, without the Mediation of anything else, by and through which their Action and Force may be conveyed from one to another, is to me so great an Absurdity that I believe no Man who has in philosophical Matters a competent Faculty of thinking can ever fall into it." (Newton, 1692–1693, p. 25)

Instead, Newton believed that

> **Q7.10**. "gravity must be caused by an Agent acting constantly according to certain laws; but whether this Agent be material or immaterial, I have left to the Consideration of my Readers." (Newton, 1692–1693, p. 25)

French scholar Pierre-Simon Laplace (1749–1827) used Michael Faraday's concept of field to express Newton's force of gravity in terms of a gravitational field. An example of a gravitational field is presented in Figure 7.1(B). The shape of the gravitational field can be viewed as a well that gets deeper and steeper as a mass such as m_1 approaches m_E. The concept of gravitational well takes on a more geometric meaning in general relativity.

In our Figure 7.1 example, two spatially separated objects with masses m_E and m_1 interact with one another when one object m_1 interacts with the field generated by the other object m_E. In this case, action at a distance between the masses is an interaction that is mediated by the field generated by object m_E. The interaction is still considered instantaneous, which became a problem after Einstein introduced the special relativistic view that nothing could travel faster than the speed of light.

In the following sections, we describe how local interactions can replace instantaneous action at a distance. The principle of locality says that object A can influence object B if the two objects can interact at a speed that does not exceed the speed of light. A local theory does not allow object A to interact with object B at a speed greater than the speed of light. By contrast, a nonlocal theory allows object A to interact with object B at a speed that exceeds light speed. Action at a distance is an example of a nonlocal theory.

7.3 Action at a Distance on the Subatomic Level

The contrast between local and nonlocal theories can be illustrated by considering the behavior of interacting particles on the subatomic level. This point of view assumes that a particle view of nature can be used to describe the subatomic realm. One of the best experimental methods for studying particle interactions is particle scattering.

Scattering occurs when the direction of a moving object is changed by interacting or colliding with another object. Scattering experiments on the subatomic level are conducted by machines that guide a collimated beam of particles into either a stationary material target or into another beam of particles. Ordinarily, the incident beam of particles must be accelerated until the incident beam attains a desired velocity. Machines built to perform particle scattering experiments are known as particle accelerators.

American physicist Richard Feynman (1918–1988) introduced spacetime diagrams in 1948 that represented interactions between particles. The basic structure of a Feynman diagram is sketched in Figure 7.2. The horizontal axis represents one space axis, and the vertical axis represents the time axis. A point on the spacetime diagram is an event. An arrow on the diagram represents the path of a particle, and the dot in Figure 7.2 is a vertex that depicts the spacetime event when an interaction occurs between two or more particles.

The cases shown in Figure 7.3 show how to interpret Feynman diagrams. Case A is a typical example of a motionless object whose clock is running. Case B is the same object with its clock running in reverse.

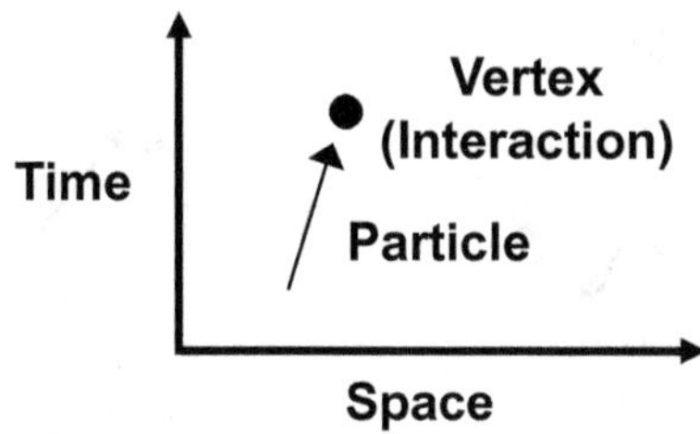

Figure 7.2. Basic Structure of a Feynman Diagram.

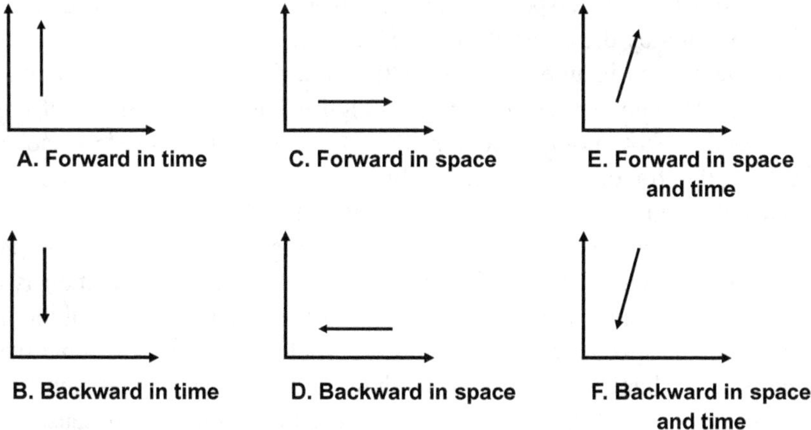

Figure 7.3. Interpreting Feynman Diagrams.

The possibility that an object can move into the past shows that the time used in Feynman diagrams is not Newtonian time, it is the time of spacetime. Cases C and D represent objects moving forward and backward in space without any passage of time. To date, there is no known physical phenomenon with the characteristics of Cases C and D. The last two cases, by contrast, are routinely observed in scattering experiments. Case E shows an object moving forward in space and time. Case F shows an object moving backward in space and time.

Feynman diagrams can be used to illustrate the interactions that were known at the beginning of the 20th century when Einstein introduced his theories of relativity. Figure 7.4 shows Feynman diagrams illustrating electromagnetic and gravitational interactions. The space and time axes are not shown but are understood to be present as in Figures 7.2 and 7.3.

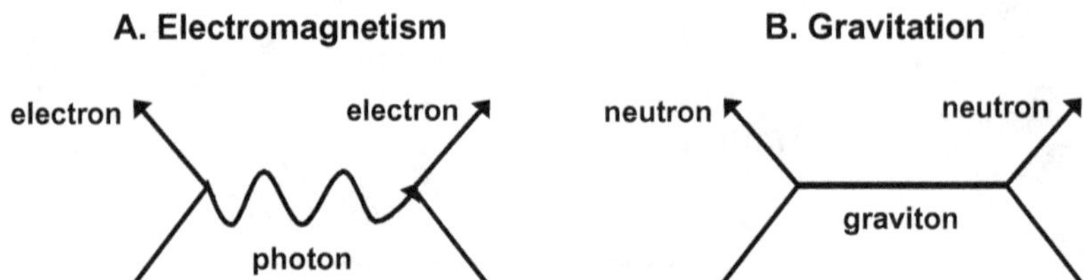

Figure 7.4. Feynman Diagrams Illustrating Electromagnetism and Gravitation.

Empirically, two electrons with the same electrical charge repel each other. From a field perspective, one electron interacts with the electric field generated by the other electron. The interaction between electrons is instantaneous and is an example of action at a distance. Another view of the interaction between two electrons in Figure 7.4(A) shows the electrons exchanging a photon or quantum of light. The photon is a massless packet of energy that travels at the speed of light. The change in the trajectory of the two electrons is a result of the exchange of a photon.

The interaction of two electrons shown in Figure 7.4(A) is an example of a local interaction. A local interaction occurs when the interaction is a result of the exchange of a particle that does not exceed the speed of light. In agreement with special relativity, massless particles such as the photon travel at the speed of light; while particles with mass travel at speeds less than the speed of light. Local interactions do not occur instantaneously because the exchanged particle does not exceed the speed of light. By contrast, instantaneous action at a distance is an example of a nonlocal interaction.

The gravitational interaction between two neutrons shown in Figure 7.4(B) is hypothesized to occur as a result of the exchange of a graviton, which has yet to be discovered. On the astronomical level, the earth's gravitational field is represented as a field. The magnitude and direction of the gravitational field generated by the earth change as the earth moves in its orbit. From the point of view of action at a distance, an object in the earth's gravitational field is instantaneously affected by the change in the gravitational field. From the point of view of locality, the change in the gravitational field must propagate through the space around the earth at a speed that does not exceed the speed of light. The propagation of the gravitational field is the basis of an experimental test of general relativity. These ideas are discussed in more detail in the next chapter.

7.4 Relationalism

Space, time, and motion are closely connected concepts. In previous chapters, we discussed a number of concepts of space, time, and motion. For example, in Chapter 2, we reviewed Aristotle's metaphysical view that the goal of every object was to attain a state of rest and unchangeability, which he called the 'prime mover'. The prime mover does not act, yet attracts everything else by its presence.

Aristotle considered spatial extension, motion, and time as continua that exist in ordered relation to one another. Motion depends on spatial extension, and time depends on motion. In this view, we observe how much time passes by observing motion or change relative to a spatial continuum. If nothing changes, time does not pass.

Aristotle's concepts of spatial extension, motion, and time were superseded by more modern ideas. In *Principia*, Isaac Newton proposed a bucket experiment to demonstrate that absolute motion could be distinguished from relative motion by observable effects (Newton, 1687, pp. 10–11). Newton's bucket experiment became a source of controversy and is discussed next.

Newton's bucket experiment

Isaac Newton used absolute space and absolute time to specify the location of objects and the duration of time. Motion could be represented by the speed of an object, which Newton defined as the ratio of a change in absolute space divided by a specified duration of absolute time.

Figure 7.5 illustrates Newton's bucket experiment. A vessel, in this case a bucket, is partially filled with water. The bucket is attached to a cord and rotated until the cord is tightly wound. Figure 7.5(A) shows the twisted cord, bucket, and water. Figure 7.5(B) shows the beginning of rotational motion of the bucket when the twisted cord is allowed to unwind. The motions of the bucket and water are summarized below for steps A through D in Figure 7.5:

A. Water is at rest relative to the bucket. The water surface is flat.
B. The bucket rotates relative to the water. The water surface remains flat initially.

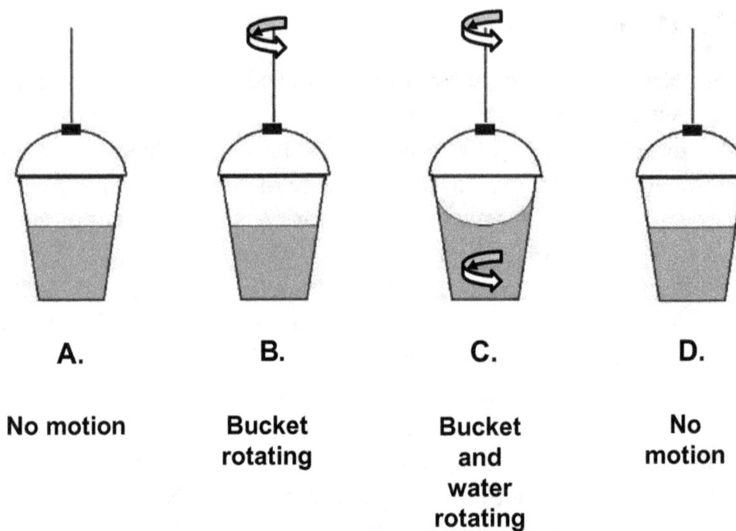

Figure 7.5. Newton's Bucket Experiment.

C. Water rotates with the bucket, and curvature appears on the water surface. The water is at rest relative to the bucket when the curvature of the water surface does not change.
D. The water surface returns to a flat surface when the bucket and water cease to rotate.

The change in the direction and speed of the rotating water shows that water is accelerating and tends to recede from the center of the bucket. Newton concluded at the end of the bucket experiment that

> **Q7.11**. "this endeavor [of the water receding from the center of the bucket] does not depend upon any translation of the water in respect to ambient bodies, nor can true circular motion be defined by such translation." (Newton, 1687, p. 11)

The Foucault pendulum

Two methods were available to measure the rotation of the earth about its polar axis before the 20th century. An astronomical method used observations of the rising and setting of distant stars to measure the revolution of

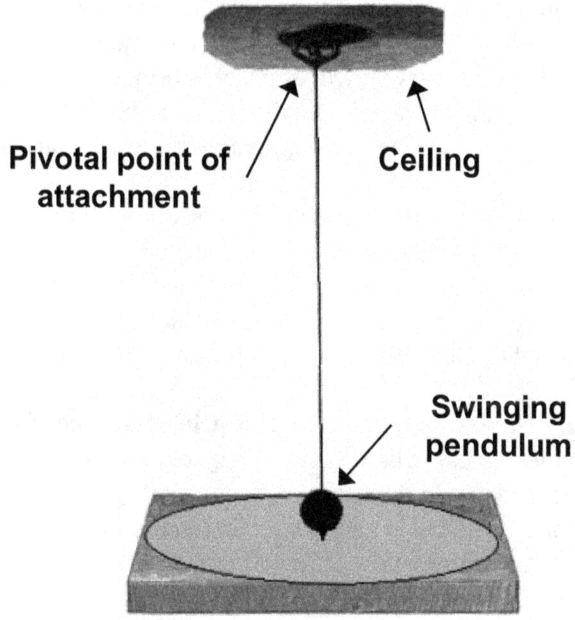

Figure 7.6. Foucault Pendulum.

the earth about its axis. French physicist Jean Foucault introduced a pendulum method in 1851. The pendulum is now known as the Foucault pendulum and is sketched in Figure 7.6. The Foucault pendulum hangs from an elevated position, such as the ceiling of a room, with a pivotal point of attachment that allows the pendulum to swing in any direction. As the earth rotates, the vertical plane of the pendulum swing gradually rotates around a vertical axis.

Mach's principle

Austrian physicist Ernst Mach (1838–1916) realized that the rotation of the earth could be observed using the earth-bound Foucault pendulum or the rotation of the earth relative to distant stars. Mach believed that the origin of inertia or mass of an object is a dynamical quantity determined by the rest of the matter in the universe. Mach introduced the hypothesis that an accelerating body experiences inertial forces that depend on the quantity and distribution of matter throughout the universe. In other words, the nonuniform motion of an object is relative motion because it depends on the quantity and distribution of matter throughout the universe. This hypothesis is known as Mach's principle.

Mach realized that Newton's analysis of the bucket experiment did not take into account ambient bodies, notably the earth and fixed stars:

> **Q7.12**. "Newton's experiment with the rotating vessel of water simply informs us, that the relative rotation of the water with respect to the sides of the vessel produces no noticeable centrifugal forces, but that such forces are produced by its relative rotation with respect to the mass of the earth and the other celestial bodies." (Mach, 1919, p. 232)

When ambient bodies are considered, Mach concluded that Newton's bucket experiment shows that water rotating relative to the Earth or fixed stars tends to recede from the center of the bucket.

The motions of the bucket and water that includes reference to 'ambient bodies' such as the earth and fixed stars are summarized below for steps A through D in Figure 7.5:

A. Water is at rest relative to the bucket, the earth, and the fixed stars. The water surface is flat.
B. The bucket rotates relative to the water, the earth, and the fixed stars. The water surface remains flat initially.
C. Water rotates with the bucket, and curvature appears on the water surface. The water is at rest relative to the bucket when the curvature of the water surface does not change. The water and bucket are rotating relative to the earth and the fixed stars.
D. The water surface returns to a flat surface when the bucket and water cease to rotate relative to the earth and fixed stars.

Mach was not the first to propose a connection between geometry and matter. A link between the motion of a body and distant stars was introduced by Anglo-Irish bishop, philosopher, and scientist George Berkeley (1685–1753). He published an essay entitled *On Motion: The Principle and Nature of Motion and the Cause of the Communication of Motions*, which is known as De Motu (Berkeley, 2012). Berkeley viewed motion as the change of place of bodies in the material world. He considered Newton's absolute space boundless but unobservable to our senses. Instead, Berkeley favored the space delineated by observable constituent bodies. This view established Berkeley as a "precursor of Mach and Einstein" (Britannica, 2022).

Relationalism and emergent properties

Gottfried Wilhelm Leibniz (1646–1716), another contemporary of Newton, also challenged the primacy of absolute space and absolute time. Leibniz proposed a relationalist view that the world of objects existed in a network of relationships. Relationalism views relationships as fundamental. The network of relationships defined distances and durations between objects and events in a relational world. From a relationalist perspective, if we removed all objects in the world, we would also remove space and time. Spatial and temporal relationships only exist between objects and events. Today, we would say that space and time are emergent properties from a network of relationships.

Emergence is an important concept in a relational world. An emergent property is a property that appears when the components of a system work together. The emergent property is not a property of any individual component. Some possible examples of emergent properties are presented in Table 7.2. The properties of solids, liquids, and gases can be considered emergent properties of collections of atoms or molecules. Space can be viewed as a property that emerges from a network of relationships between objects. Time can be viewed as a way to organize change.

A practical application of replacing a relational world with a mathematical representation is illustrated in Figure 7.7. Figure 7.7(A) shows a rock wall in the foothills of the Rocky Mountains near Denver, Colorado. A mathematical model of a segment of the rock wall can be constructed by overlaying a mathematical grid over the segment of interest as shown in Figure 7.7(B). The mathematical grid is designed to model the

Table 7.2. Possible Examples of Emergent Properties.

Emergent Property	Relational World
Phase (solid, liquid, or gas)	Collection of atoms and molecules
Space	Network of spatial relationships between objects and events
Time	Network of temporal relationships between objects and events

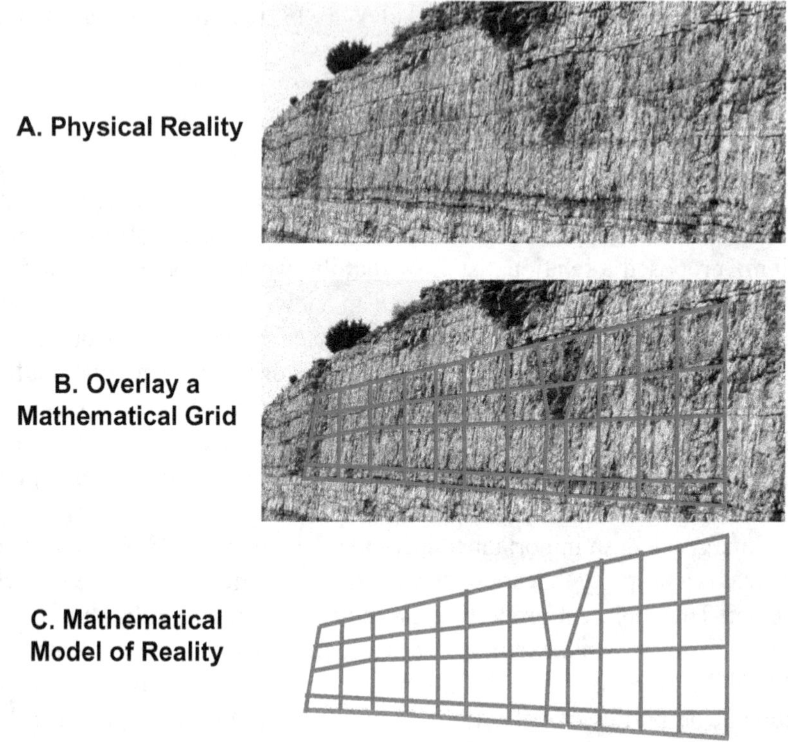

Figure 7.7. Mathematical Representation of Physical Reality.

distribution of matter in the chosen segment of the rock wall and maintain relationships between different parts of the rock wall. The mathematical model in Figure 7.7(C) can be said to emerge from the physical reality in Figure 7.7(A). Different gridding schemes could be used to model the underlying physical reality. The physical reality does not change when a different grid is used, but the representation of the physical reality does change.

# # #

The special theory of relativity replaced absolute space and absolute time with a spacetime continuum that considered space and time as relative concepts. The spacetime continuum can be considered an emergent property of a relational view of space and time. Einstein knew that he had to extend the mechanics of special relativity from inertial systems to accelerating systems. Mach's principle — the idea that inertial forces acting on an accelerating body were due to the distribution and quantity of matter in the universe — helped Einstein formulate the general theory of relativity because it suggested a connection between matter and the geometry of the universe (Britannica, 2021).

Endnote

1. Sources about gravity include Misner *et al.* (1973), Wheeler (1990), Peebles (1993), Aughton (2008), Weinberg (2008), Fraknoi *et al.* (2018), Bennett *et al.* (2018), Teerikorpi *et al.* (2019), Malkan and Zuckerman (2020), and Hartle (2021).

Chapter 8

General Relativity

German physicist Albert Einstein (1879–1955) was unable to present a theory of relativity in 1905 for reference frames which were accelerating with respect to one another. After a decade of research, Einstein (1916) published a relativity theory for accelerating reference frames that he called the general theory of relativity because it applied to observers in either uniform or accelerating motion with respect to one another. A year later Einstein (1917) published an article that explored cosmological models within the context of the general theory of relativity. We begin our discussion of general relativity by reviewing the intellectual trail followed by Einstein.[1]

8.1 Deflection of Light Rays

The year 1905 is sometimes known as Albert Einstein's miracle year. In addition to the special theory of relativity and the equivalence between mass and energy, Einstein published papers on Brownian motion and the photoelectric effect. His 1905 papers have been collected and published by John Stachel (Stachel, 1998).

Einstein's paper on Brownian motion was entitled *On the Movement of Small Particles Suspended in Stationary Liquids Required by the Molecular-Kinetic Theory of Heat*. At the time, the view that matter was composed of atoms was still being questioned. Brownian motion is the random motion of microscopic particles suspended in a fluid. The motion is attributed to collisions with atoms or molecules in the surrounding fluid. It was empirical evidence for the atomic theory of matter.

Einstein provided a statistical analysis of the behavior of microscopic particles in the presence of a fluid of colliding atoms and molecules. French physicist Jean-Baptiste Perrin (1870–1942) relied on Einstein's 1905 theory of Brownian motion to carry out "the first modern determination" (Lautrop, 2005, p. 4) of Avogadro's number from Brownian motion in 1908. Avogadro's number is the number of atoms in 12 g of carbon-12, the carbon isotope with 6 protons and 6 neutrons. According to Benny Lautrop, Perrin's experiments "were not only seen as a confirmation of [Einstein's] theory but also as the most direct evidence for the reality of atoms and molecules" (Lautrop, 2005, p. 5). Perrin received the Nobel Prize in Physics in 1926 for his work on Brownian motion.

The photoelectric effect was the observation that a beam of light with a suitable frequency could cause the ejection of electrons from a material. Attempts to explain the photoelectric effect using the then widely accepted wave theory of light were not successful. Einstein proposed an alternative explanation in his 1905 paper entitled *On a Heuristic Point of View Concerning the Production and Transformation of Light*. He suggested that light could be represented as a discrete packet, or quantum, of energy:

> **Q8.1**. "According to the assumption considered here, in the propagation of a light ray emitted from a point source, the energy is not distributed continuously over ever-increasing volumes of space, but consists of a finite number of energy quanta localized at points of space that move without dividing, and can be absorbed or generated only as complete units." (Einstein, 1905c, p. 178)

A quantum of light energy can be illustrated as a packet or bundle of energy, as illustrated in Figure 8.1. The quantum of light energy is now called a photon. Einstein received the Nobel Prize for Physics in 1921 for his theory of the photoelectric effect. It inspired the development of the

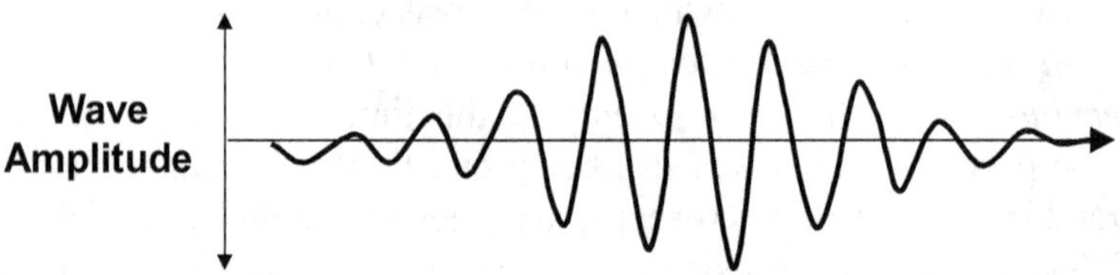

Figure 8.1. Sketch of a Wave Packet.

concept of wave–particle duality in which a quantum system could behave as a wave or a particle. The wave–particle duality is discussed in Chapter 11.

Einstein postulated that the trajectory of a ray of light could be altered by an observable amount if the gravitational field of a star was strong enough. From a more modern perspective, the energy content of a quantum of light can be influenced by gravity. Einstein predicted the deflection of light rays by the gravitational field of the Sun in 1911:

> **Q8.2**. "Rays of light, passing close to the Sun, are deflected by its gravitational field." (Einstein, 1911, p. 99)

He then suggested a practical experiment:

> **Q8.3**. "As the fixed stars in the parts of the sky near the Sun are visible during total eclipses of the Sun, this consequence of the theory may be compared with experience... It would be a most desirable thing if astronomers take up the question here raised." (Einstein, 1911, p. 108)

Figure 8.2 shows a star on the side of the Sun opposite to the Earth. Light rays from the star are deflected by the gravitational field of the Sun. Figure 8.2(A) shows the apparent position of a star based on the deflected light ray, and Figure 8.2(B) shows the actual position of the star.

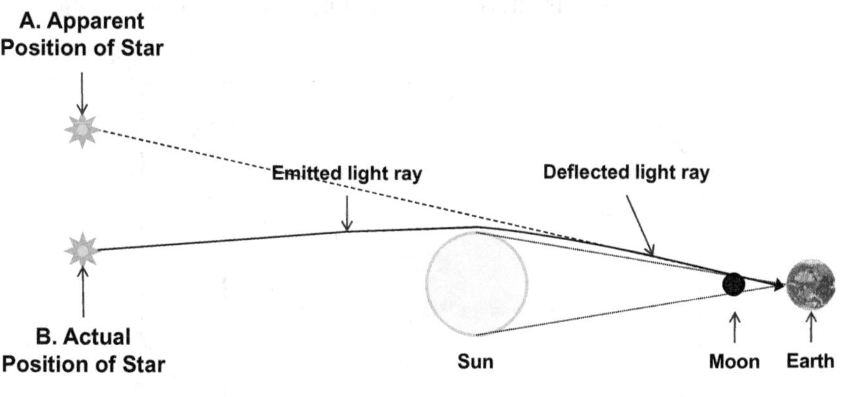

Figure 8.2. Deflection of Light Rays by the Sun.

English astronomer and physicist Arthur S. Eddington (1882–1944) first confirmed the deflection of light rays during a total solar eclipse in 1919. Total solar eclipses occur when the Moon passes between the Sun and Earth. The Moon blocks the glare of the Sun and makes it possible to view stars from Earth that are otherwise concealed by the Sun's glare. Subsequent experiments have further corroborated Einstein's prediction of the deflection of light rays by a gravitational field. The deflection of light rays is the basis of gravitational lensing. A celestial object that can be used to bend and focus light from a more distant object is an example of a gravitational lens.

8.2 Frames of Reference and the Elevator Gedanken Experiment

In the period following the publication of his special theory of relativity (Einstein, 1905a) and prior to the publication of his general theory of relativity (Einstein, 1916), Einstein constructed gedanken, or thought, experiments to help guide his development of relativity. Here we consider an elevator gedanken experiment with three scenarios.

Suppose an observer is standing on a scale in a windowless elevator. The elevator can be considered a frame of reference (FoR). An FoR is needed to analyze motion. A basic FoR consists of a set of coordinate axes to label points in space and a clock to track time, as illustrated in Figure 8.3. FoRs can be inertial or accelerating. An inertial FoR in Newtonian physics is at rest or moving with a constant velocity.

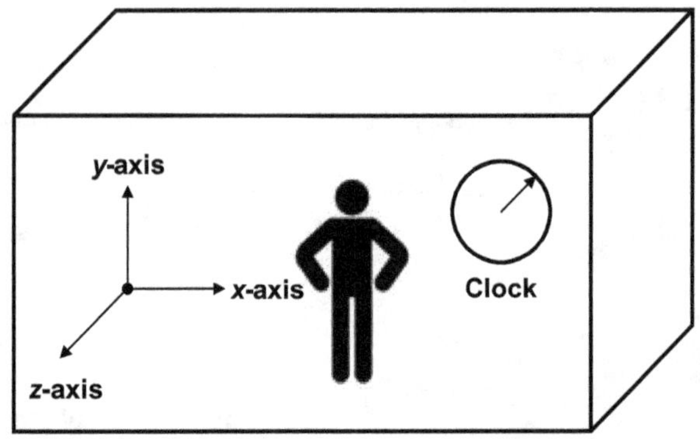

Figure 8.3. Basic FoR for an Observer in a Room.

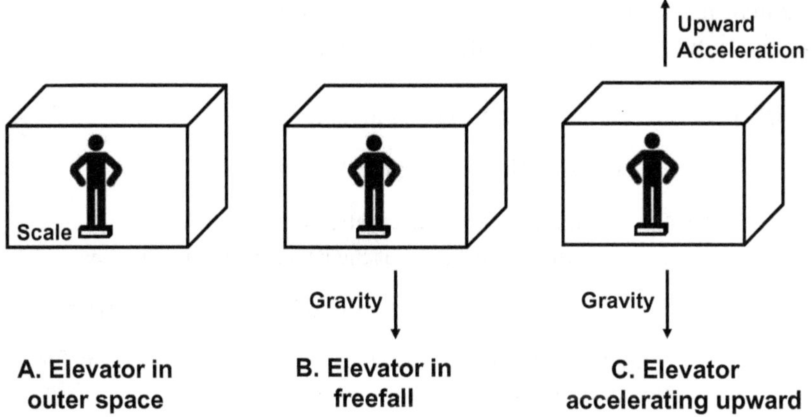

Figure 8.4. Scenarios of the Elevator Gedanken Experiment.

Any frame that accelerates with respect to an inertial frame is known as an accelerating FoR or non-inertial FoR.

The scenarios of the elevator gedanken experiment are illustrated in Figure 8.4. In Scenario A, the elevator is in outer space and is not in the presence of a gravitational field. Consequently, the observer does not exert any pressure on the scale and is weightless. In Scenario B, the elevator is in free fall in a gravitational field. Both the observer and the scale are also in free fall in this scenario. The observer does not exert any pressure on the scale and is weightless. In Scenario C, the elevator is accelerating upward relative to a gravitational field. In this case, the observer exerts pressure on the scale and has weight.

When the elevator FoR is in free fall in Scenario B, it behaves as if the gravitational field was not present. On the other hand, when the elevator is accelerating upward relative to a gravitational field in Scenario C, the weight of the observer in the elevator depends on the net acceleration. Einstein used the results of such gedanken experiments and his understanding of nature to propose an equivalence principle: an observer in an inertial FoR is not able to distinguish between free fall and lack of gravity. The equivalence principle is discussed in more detail next.

8.3 Einstein's Principle of Equivalence

Albert Einstein published a book for laypeople in 1916 that was subsequently revised in 1924 (Einstein, 1924). Einstein included the following argument for the equivalence principle in the book. He began by drawing a distinction between inertial mass and gravitational mass:

> **Q8.4**. "According to Newton's law of motion, we have
>
> $$(\text{Force}) = (\text{Inertial mass}) \times (\text{Acceleration}),$$
>
> where the 'inertial mass' is a characteristic constant of the accelerated body. If now gravitation is the cause of the acceleration, we then have
>
> $$(\text{Force}) = (\text{Gravitational mass}) \times (\text{Intensity of the gravitational field}),$$
>
> where the 'gravitational mass' is likewise a characteristic constant for the body." (Einstein, 1924, p. 59)

Inertial mass measures the tendency of an object to continue in a straight line unless acted upon by an external force. On the other hand, gravitational mass is a measure of the gravitational strength of one object relative to another.

Einstein combined the two force equations in quote Q8.4 and found that

$$(\text{Acceleration}) = \frac{(\text{Gravitational mass})}{(\text{Inertial mass})} \times (\text{Intensity of the gravitational field}).$$

He reasoned that if acceleration is to be

> **Q8.5**. "independent of the nature and the condition of the body and always the same for a given gravitational field, then the ratio of the gravitational to the inertial mass must likewise be the same for all bodies. By a suitable choice of units, we can thus make this ratio equal to unity $\left[\text{that is, } \frac{(\text{Gravitational mass})}{(\text{Inertial mass})} = 1\right]$." (Einstein, 1924, p. 59)

This conclusion justified the Principle of Equivalence, which says that the gravitational mass of a body is equal to its inertial mass within an FoR.

Hungarian physicist Loránd Roland, baron von Eötvös (1848–1919) showed the equivalence of inertial mass and gravitational mass using a torsion balance. Key elements of Eötvös's torsion balance are sketched in Figure 8.5. Wire 1 is suspended at its upper end. The lower end of

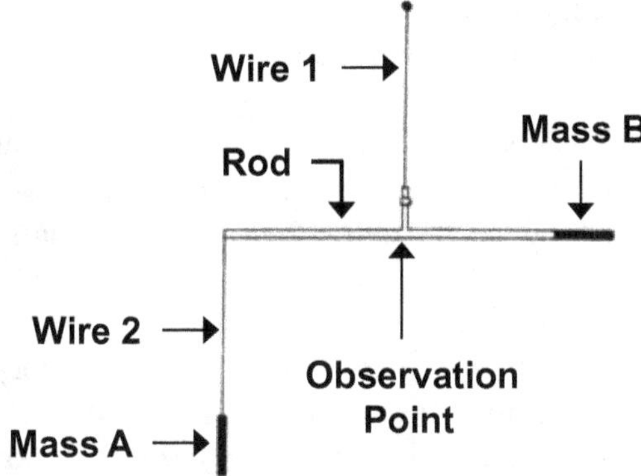

Figure 8.5. Key Elements of Eötvös's Torsion Balance.

wire 1 is attached to a rod. The left-hand side of the rod supports mass A suspended by wire 2, while an equivalent mass B is attached to the right-hand side of the rod. Wire 1 is the torsion wire, and the rod is the balance bar. The system consisting of wires, masses, and rod is enclosed in walls that insulate the system from temperature variations or air disturbances.

The rod would be balanced in a gravitational field if masses A and B were at the same elevation and equidistant from wire 1. The wire supporting the rod in Figure 8.5 is subjected to a torsion (twist) because of the gravitational imbalance caused by the difference in elevation of mass A and mass B. The twisting force is related to the gravitational force, and analysis of forces shows that inertial mass is equivalent to gravitational mass.

Many experiments have improved on the results of Eötvös's torsion balance and validated Einstein's Principle of Equivalence.

8.4 Einstein's General Theory of Relativity

Albert Einstein knew that the special theory of relativity was too limited because it only applied to inertial FoRs. To extend the special theory, Einstein proposed a more general perspective:

> **Q8.6.** "The laws of physics must be of such nature that they apply to systems of reference in any kind of motion. Along this road we arrive at an extension of the postulate of relativity." (Einstein, 1916, p. 113)

The general theory of relativity should apply to a broad range of reference systems. In addition, Einstein believed

> **Q8.7**. "the general laws of nature are to be expressed by equations which hold good for all systems of co-ordinates, that is, are co-variant with respect to any substitutions whatever (generally co-variant)." (Einstein, 1916, p. 116)

Einstein realized that the spacetime of the special theory of relativity, Minkowski spacetime, must be too restrictive. Minkowski spacetime is an example of Euclidean geometry. An alternative geometry is non-Euclidean geometry. The primary difference between the geometry of Euclid, Euclidean geometry, and non-Euclidean geometry is the relationship between parallel lines. Figure 8.6 illustrates the difference between Euclidean and non-Euclidean geometry on a two-dimensional surface.

Figure 8.6(A) shows two straight lines (a, b) that are perpendicular to a third line (c). If the two straight lines (a, b) never intersect when both lines are extended without limit, the two-dimensional surface in Figure 8.6(A) is Euclidean. The distance between straight lines (a, b) never changes. By contrast, both Figures 8.6(B) and 8.6(C) show non-Euclidean examples where the distance between lines (a, b) does change. Figure 8.6(B) is an example of a hyperbolic non-Euclidean geometry where the distance between lines (a, b) increases as lines (a, b) move away from the point of intersection with the perpendicular line (c). Figure 8.6(C) is an example of an elliptic non-Euclidean geometry where the distance between lines (a, b) decreases as we move away from the point of intersection with perpendicular line (c).

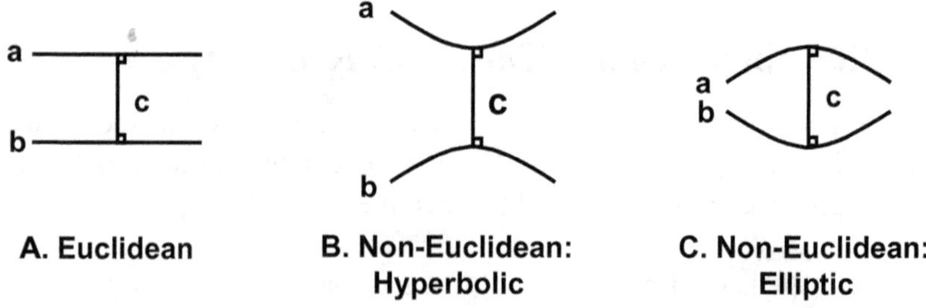

Figure 8.6. Euclidean and Non-Euclidean Geometry.

Gedanken experiments with free-falling objects in a gravitational field and the Principle of Equivalence helped guide Einstein to the realization that moving objects may only be following a path in spacetime. The spacetime was not flat, however, but curved. He adopted a geometric interpretation of gravity as the geometry of curved spacetime. This interpretation was succinctly expressed by American physicist John Archibald Wheeler (1911–2008): "Spacetime tells matter how to move; matter tells spacetime how to curve" (Wheeler, 2010, p. 235). A region of spacetime is curved by the mass and energy within it. If there is no mass or energy in the spacetime region, the spacetime region is flat.

Curved spacetime results from the presence of mass and energy in an otherwise flat spacetime region. Einstein hypothesized that curved spacetime can be mathematically described by non-Euclidean geometry. From the geometric perspective, flat spacetime is described by Euclidean geometry. The general theory of relativity corresponds to the special theory of relativity when gravity is absent or negligible (Einstein, 1916, p. 157). The special theory of relativity applies to spacetime regions with flat spacetime.

8.5 Early Tests of General Relativity

Albert Einstein highlighted three experimental tests of general relativity (Einstein, 1924, Appendix C). The deflection of light rays by the Sun, the advance of the perihelion of the planet Mercury, and the gravitational redshift. The deflection of light rays by the Sun was discussed above. The advance of the perihelion of Mercury and the gravitational redshift are discussed here.

The perihelion of Mercury

The perihelion of a planet tracing an elliptical orbit around a spherical star is the point where the planet is closest to the star. Figure 8.7(A) illustrates the perihelion of Mercury as it orbits the Sun. The perihelion of Mercury is significant as a test of general relativity because the perihelion changes as the planet orbits the Sun. The change in perihelion is called the advance of perihelion or the precession of perihelion of Mercury. It is sketched in Figure 8.7(B). Newtonian mechanics did not accurately predict the observed advance of the perihelion of Mercury. Einstein confirmed

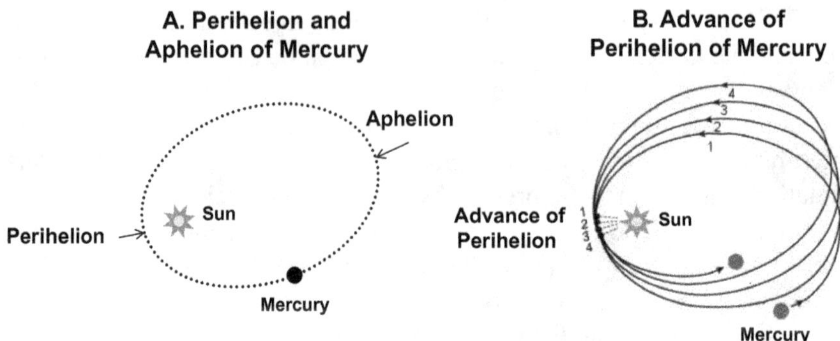

Figure 8.7. Advance of the Perihelion of Mercury.

general relativity by correctly calculating that the perihelion of Mercury advances 43 arcsec per century.

Gravitational redshift and blueshift

The gravitational redshift refers to the change in frequency that occurs when photons of light move from a larger gravitational field to a smaller gravitational field. This can occur when a photon of light is emitted by a star. The minimum speed needed for an object such as a photon to escape the gravitational pull of a massive body such as a star is the escape velocity. If a sufficiently dense star has an escape velocity that exceeds the speed of light, the star is considered a black hole because light cannot escape the gravitational pull of the star. From a geometric perspective, the curvature of spacetime is so great that it creates a gravitational well that is too steep for the photon to escape.

Another possibility occurs when a photon is emitted from the surface of a star and the speed of light exceeds the escape velocity of the star. The photon can now expend energy to escape from the gravitational pull of the star. The speed of the photon cannot change, however, so the decrease of photon energy is achieved by decreasing the frequency of the photon. A decrease in photon frequency corresponds to an increase in photon wavelength. The frequency shifts to the red end of the electromagnetic spectrum and is called a redshift. It is worth noting that the frequency of a photon falling into a gravitational well increases and shifts to the blue end of the electromagnetic spectrum as it gains energy.

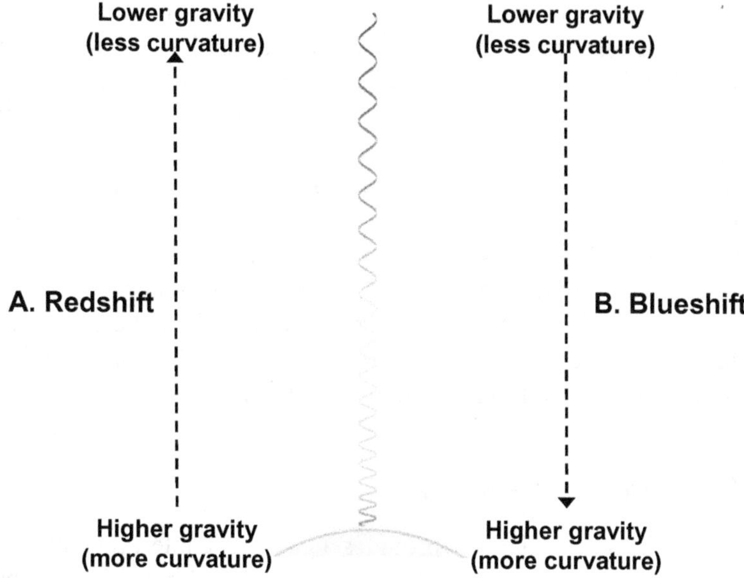

Figure 8.8. Gravitational Redshift and Blueshift.

Figure 8.8 illustrates the gravitational redshift and blueshift. The redshift in Figure 8.8(A) occurs when the photon travels away from the star. The gravitational field decreases, which corresponds to a decrease in the curvature of spacetime. The frequency of the photon decreases as its wavelength increases. The blueshift in Figure 8.8(B) occurs when the photon travels toward the star. The gravitational field increases, which corresponds to an increase in the curvature of spacetime. The frequency of the photon increases as its wavelength decreases.

A time dilation effect can be associated with the gravitational frequency shift of light by recognizing that a clock runs slower as it approaches the mass generating the gravity. A clock runs faster in low gravity than it does in high gravity. Therefore, clock C_L in a low gravity FoR runs faster than clock C_H in a high gravity FoR.

Einstein first introduced the idea that the frequency of light could be shifted by gravity in 1907. He revisited the topic on several occasions (Earman and Glymour, 1980).

The gravitational redshift has been observed in the laboratory and in astrophysical measurements. The related time dilation effect has been verified by flying atomic clocks at different altitudes on Earth.[2] An atomic clock keeps time by measuring the electromagnetic signal that is emitted when an atomic electron transitions from one energy level to another. The spectrum of emitted light is used as a frequency standard for keeping time. The emission of light from an atom is a quantum effect that is discussed in more detail in a later chapter.

Atomic clocks are accurate standards of time and frequency. They are used in satellite navigation systems such as the Global Positioning System (GPS). Satellite navigation systems rely on atomic clocks aboard orbiting satellites and Earth-bound reference clocks. A comparison of the time and frequency observations made by orbiting and Earth-bound atomic clocks must take into account relativistic effects associated with the relative motion of the clocks and time dilation in a gravitational field.

8.6 Gravitational Waves

The general theory of relativity provides a geometric interpretation of the gravitational interaction and predicts, among other things, the existence of gravitational waves and black holes. Black holes were introduced above. Gravitational waves[2] are disturbances of spacetime curvature that propagate through space at the speed of light. Detectable gravitational waves can be generated by the collision of two stars.

Gravitational waves generated by two black holes spiraling into each other were detected in 2015 by Laser Interferometer Gravitational-Wave Observatories (LIGO) in Louisiana and Washington. Gravitational-wave detectors consist of two intersecting, perpendicular arms with mirrors at each end. A beam of laser light from the end of one arm is split into two beams that travel along the perpendicular arms. The light from the beams is reflected by mirrors. The recombined beams of light form an interference pattern when a passing gravitational wave alters the length of the perpendicular arms.

# # #

The scientific view of space, time, and gravity changed significantly during the years between Isaac Newton and Albert Einstein. Differences between the gravitational theories of Newton and Einstein depend on the strength of the gravitational field. The absence of gravity corresponds to flat spacetime. Newton's theory of gravity is an approximation of

Einstein's general theory of relativity when gravity is weak and the curvature of spacetime is small. General relativity is applicable when gravity is strong and the curvature of spacetime is large. These conditions occur in modern cosmology, which is introduced in the next chapter.

Endnotes

1. The literature on relativity is extensive. See, for example, Einstein (1924), Pauli (1958), Misner *et al.* (1973), Brouwer (1980), Pais (1982), Jackson (1987), Peebles (1993), Goldsmith (1995), Liddle (1999), Ludvigsen (1999), Bambi (2018), Teerikorpi *et al.* (2019), Grøn (2020), and Hartle (2021).
2. Sources on tests of relativity include Kisslinger (2017), Teerikorpi *et al.* (2019), Grøn (2020), Hartle (2021), and Perkowitz (2021).

Part 3

Cosmology and the Inflationary Universe

Chapter 9

Introduction to Modern Cosmology

Modern cosmology is based on the application of Einstein's general theory of relativity (Einstein, 1916). A set of solutions of the field equations of general relativity that are applicable on a universal scale can be considered a cosmological model. Different solutions of the field equations could be obtained by specifying different assumptions. The validity of a cosmological model depends on a comparison of the cosmological model with physical evidence. The comparison is a means of validating, or invalidating, the assumptions underlying each cosmological model.

Modern cosmological models highlight how the role of time has changed from Isaac Newton's absolute time to Einstein's geometric spacetime on a universal scale. Pioneering work in modern cosmology is discussed in this chapter.[1]

9.1 Einstein's Cosmology

Albert Einstein first applied the field equations of general relativity to cosmology in 1917. Einstein reminded the reader that the geometry of the universe depended on the distribution of mass–energy in the universe:

> **Q9.1**. "According to the general theory of relativity the metrical character (curvature) of the four-dimensional spacetime continuum is defined at every point by the matter at that point and the state of that matter." (Einstein, 1917, p. 180)

He assumed that the universe was finite:

> **Q9.2**. "The universal continuum in respect of its spatial dimensions is to be viewed as a self-contained continuum of finite spatial (three-dimensional) volume." (Einstein, 1917, p. 180)

His cosmological model showed that

> **Q9.3**. "the curvature of space is variable in time and place, according to the distribution of matter, but we may roughly approximate to it by means of a spherical space." (Einstein, 1917, p. 188)

Einstein believed his cosmological model was logically consistent but acknowledged that he had to make a change to his theory:

> **Q9.4**. "In order to arrive at this consistent view, we admittedly had to introduce an extension of the field equations of gravitation which is not justified by our actual knowledge of gravitation." (Einstein, 1917, p. 188)

The extension of the field equations was a new term that was added

> **Q9.5**. "only for this purpose of making possible a quasi-static distribution of matter, as required by the fact of the small velocities of the stars." (Einstein, 1917, p. 188)

The term he added is now known as the cosmological constant.

Einstein's cosmological model represented a static, very slowly changing, or quasi-static universe. By the end of the 1920s, astronomical observations provided evidence of an expanding universe, as we discuss below. Einstein learned that adding a term to his field equations to construct a quasi-static universe was incorrect. It is now believed that the cosmological constant can be used to represent such phenomena as vacuum energy, dark mass, and dark energy. American physicist Steven Weinberg concluded that

> **Q9.6**. "Einstein's mistake was not that he introduced the cosmological constant — it was that he thought it was a mistake." (Weinberg, 2008, p. 45)

9.2 Cosmological Models Before 1929

Dutch astronomer Willem de Sitter (1872–1934) was a contemporary of Einstein. He began working with Einstein's general relativity in 1916 and published a solution of the field equations in 1917 (de Sitter, 1917a, 1917b). His solution was a model of an expanding universe devoid of matter except for one test-body:

> **Q9.7**. "If all matter were destroyed, with the exception of one material particle, then would this particle have inertia or not?" (de Sitter, 1917b, p. 5)

De Sitter questioned the view associated with Mach's principle that the inertia of a material body was due to the quantity and distribution of matter elsewhere in the universe. The Machian view implies that inertia does not exist in a world without matter. De Sitter considered an alternative point of view. He introduced the term 'world-matter' and suggested that world-matter is all of the gravitating matter in the universe. From this point of view:

> **Q9.8**. "'Inertia' is produced by the whole of the world-matter, and 'gravitation' by its local deviations from homogeneity." (de Sitter, 1917b, p. 5)

Evidence at the time favored a static universe, so de Sitter's expanding universe model was met with skepticism. De Sitter found a static solution that was more acceptable to his contemporaries. His static solution was based on the assumption that the components of the metric tensor did not depend on time (Ferrarese, 2012). The metric tensor is a mathematical function that can be used to calculate the distance between two points in a geometric space such as spacetime.

In the 1920s, Russian mathematician Alexander Friedmann (1888–1925) and Belgian cosmologist and priest Georges Lemaître (1894–1966) found dynamic solutions of Einstein's unmodified field equations, that is, the field equations without the cosmological term. Friedmann solved Einstein's field equations in 1922 for a model of a dynamic universe, that is, a universe that could change with time. His model assumed the universe was homogeneous and isotropic. Physical quantities are the same at

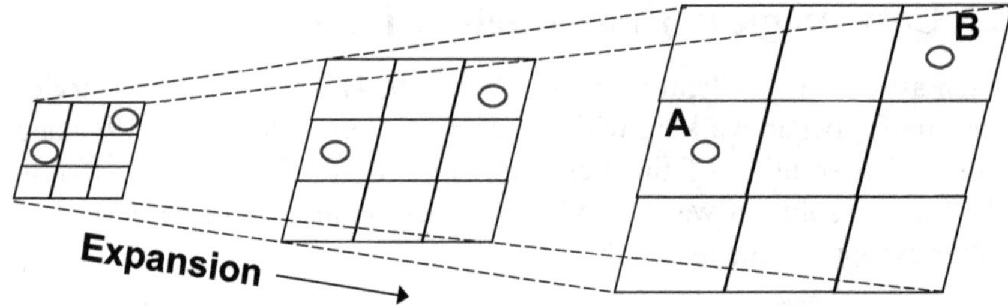

Figure 9.1. Expansion of a Coordinate Grid.

every point in space in a homogeneous universe. The universe is isotropic if it looks the same in every direction. The assumption of homogeneity and isotropy is known as the cosmological principle.

Friedmann introduced a cosmic scale factor to represent the expansion of the universe (Odenwald, 2019, Chapter 4; Liddle, 1999, Chapter 3). Suppose every galaxy is located at a fixed position on a grid. The distance between the galaxies increases as the cosmic scale factor $a(t)$ increases with time t, but the size of a galaxy does not change because the local mass–energy density is significantly larger than the corresponding density on the scale where the homogeneity assumption is applicable. Figure 9.1 illustrates the effect of the cosmic scale factor on an expanding grid with two galaxies A and B.

Lemaître independently developed a dynamic universe model from Einstein's field equations in 1927. Lemaître combined general relativity with recent astronomical observations to develop a model universe that could represent a static, expanding, or contracting universe. He showed that the velocity of a galaxy receding from an observer in an expanding universe is proportional to its distance from the observer.

The Friedmann and Lemaître models described an expanding universe rather than a static universe like the static solutions published by Einstein and de Sitter. Evidence that the universe was expanding was beginning to appear.

9.3 Early Evidence of an Expanding Universe

American astronomer Edwin Hubble (1889–1953) published evidence of an expanding universe in 1929. He relied on his observations and observations conducted by colleagues, notably American astronomer Vesto

Slipher (1875–1969). Slipher compared light from distant galaxies (then considered nebulae) with light from nearer, similar galaxies and found a shift in atomic spectra. The shift in the spectra from galaxies is a Doppler shift. It is discussed in more detail here, followed by a discussion of the discovery of Hubble's law.

Cosmological redshift

One of the tools that scientists use is the cosmological redshift, which is an optical version of the Doppler effect for sound. Sound is the wavelike transmission of vibrational energy. Since it is wavelike, sound has a wavelength and frequency (or pitch). If the wavelength of sound changes with respect to an observer, so also will its pitch. As an illustration, imagine standing next to a road as a car approaches an intersection (Figure 9.2).

The driver honks the horn as the car enters the intersection. The sound of the horn continues until the car exits the crossing. To the driver, the sound of the horn never changes because the driver is moving with the car. To an observer standing next to the intersection, the sound of the horn rises in pitch as the car approaches, and then falls as the car recedes. This change in pitch was first explained by Austrian physicist and mathematician Christian Andreas Doppler (1803–1853).

The driver of the car does not hear a change in pitch because the sound of the horn and the driver are stationary relative to each other. The change in pitch we hear is the Doppler effect for sound. As the car approaches the intersection, the number of sound waves reaching us in a given time interval increases. This is the cause of the increase in pitch we

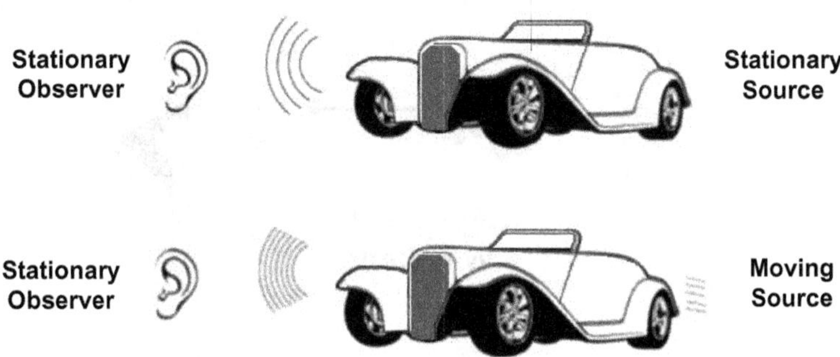

Figure 9.2. Doppler Effect for Sound.

hear. When the car passes and recedes, the number of waves reaching us in a given time interval decreases so that we hear a decreasing pitch. Changes in wavelength or pitch can result from the relative motion of the source of wave motion and an observer.

The Doppler effect applies to any wave phenomenon. It states that the frequency, or equivalently, the wavelength, of a moving wave depends on the motion of the observer. Suppose we consider a source of light, such as a star, that is being observed by a stationary observer. If the star is not moving relative to the observer, the wavelength of the emitted light is illustrated in Figure 9.3(A). If the source of a wave of light is receding from a stationary observer, the wavelength of waves passing the observer will increase, as shown in Figure 9.3(B). The increase in wavelength associated with a receding light source is known as a redshift. By contrast, if the source of a wave of light is approaching a stationary observer, the

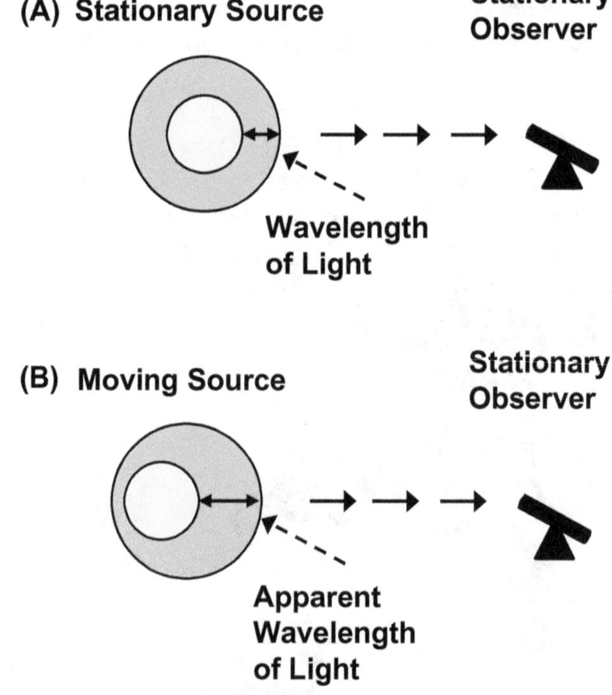

Figure 9.3. (A) Stationary Light Source and Stationary Observer. (B) Receding Light Source and Stationary Observer.

wavelength of waves passing the observer will decrease. In the case of an approaching light source, the decrease in wavelength is known as a blueshift.

The change in the wavelength of waves emitted by a moving source is considered a shift. The shift can be detected by examining the spectrum of light emitted by known elements. Many mechanisms at work in a star can radiate light. The mechanism of interest to us is the release of energy from an atom that has absorbed energy from the star. The energetically excited atom relaxes to a less energetic state by releasing energy in the form of an electromagnetic wave (light). Each kind of atom generates a spectrum, or a distribution of wavelengths, that is an identifying characteristic of that atom. For example, if we measure the spectra of hydrogen and helium in the laboratory, the spectrum of one hydrogen atom is the same as the spectrum of another hydrogen atom but differs from the helium spectrum. By determining the spectra of atoms on Earth, we can identify the types of atoms present in stars by comparing their spectra.

We can compare the absorption line spectrum obtained in a laboratory with the spectrum of light emitted by the source. Figure 9.4 compares the laboratory absorption line spectrum, which we can consider a stationary observer, with the spectra associated with approaching and receding galaxies. The three dark absorption lines are shifted to the red (redshift) when the source is receding from the laboratory, while they are shifted to the blue (blueshift) when the source is approaching. Scientists measured absorption line spectra of elements like hydrogen and helium and noticed that the absorption lines were redshifted, which implied the source was receding, or moving away, from Earth.

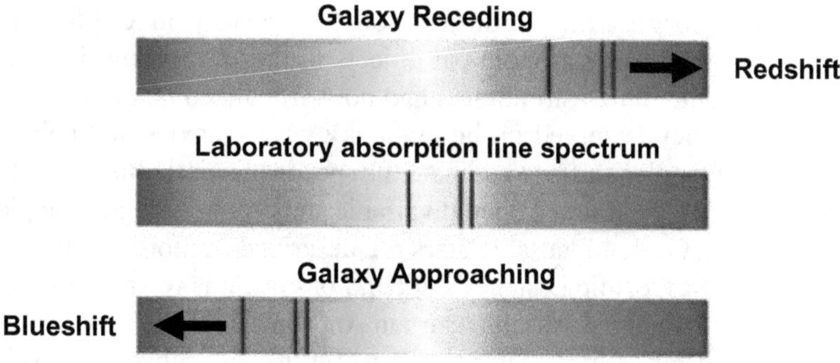

Figure 9.4. Doppler Effect and Absorption Line Spectrum.

Vesto Slipher began his professional career in astronomy in 1901 when he took a position at the Lowell Observatory in Flagstaff, Arizona (Clark, 2021). Slipher worked for American businessman and astronomer Percival Lowell (1855–1916), Founder and Director of the observatory. Lowell's work mapping the surface of Mars at the end of the 19th century seemed to show the presence of Martian canals and popularized the idea of intelligent life on Mars. This helped encourage the growth of science fiction, notably *The War of the Worlds* by H.G. Wells (Wells, 1898).

Slipher initially assisted Lowell in the study of the surface of Mars. In 1912, Lowell asked Slipher to use the observatory's 24-in refracting telescope to measure Doppler shifts of prominent spiral objects, which were then considered nebulae or possibly island universes. Astronomers expected the spiral objects to be moving toward and away from Earth in approximately equal numbers. Slipher first studied the Andromeda Galaxy, a spiral object originally known as the Andromeda Nebula, and found that it was moving toward Earth at approximately 300 km/s. He based his calculations on the Doppler shift in the spectra from the spiral objects (Slipher, 1915; Nussbaumer, 2014). The speed of Andromeda was so large that it implied that Andromeda was not in the Milky Way Galaxy. When he studied more spiral objects, he found that almost all of the spiral objects were moving away from Earth. Slipher reported that 12 out of the 15 spiral objects he studied were moving away from Earth (Slipher, 1915).

9.4 Hubble's Law

The realization that the universe contained more than one galaxy occurred early in the 20th century. Astronomer Edwin P. Hubble (1889–1953) used a 100-in telescope at the Mount Wilson Observatory in California to observe the motion of astronomical objects in the sky during the 1920s and 1930s. At the time, astronomers had not yet realized that many of the spiral objects they observed in the night sky were galaxies, not nebulae. Hubble used Doppler shifts to estimate the speeds of spiral objects relative to Earth. In 1924, he found Cepheid variable stars on a photographic plate of Andromeda. Cepheid variable stars are described in more detail in our discussion of the Cosmic Distance Ladder. For now, it is worth noting that the presence of Cepheid variable stars in Andromeda made it possible to measure the distance from Andromeda to Earth. He realized that spiral objects like Andromeda were galaxies outside of the Milky Way (Ventrudo, 2012).

Hubble prepared a diagram showing the radial velocity of a galaxy in kilometers per second (km/s) plotted against the distance from Earth to that galaxy in megaparsecs (Mpc). The radial velocity of the galaxy with respect to Earth is the rate of change of the distance between the galaxy and Earth. A megaparsec is approximately 3.26 million light years, and a light year is the distance light travels in a year (approximately 9.5 trillion km or almost 5.9 trillion miles). The plot of radial velocity versus distance is now called the Hubble diagram. Examples are shown in Figure 9.5. The linear correlation that shows that galaxies farther from Earth are moving away from us at a greater velocity than galaxies closer to Earth is called Hubble's law. It is sometimes called the Hubble–Lemaître law because Lemaître predicted the velocity–distance relationship (Lemaître, 1917).

Figure 9.5(A) shows Hubble's velocity–distance relation using data from his 1929 paper (Hubble, 1929). Figure 9.5(B) shows Hubble and Humason's velocity–distance relation using data from their 1931 paper (Hubble and Humason, 1931). The dots at the lower left of Figure 9.5(B) are data points from Hubble's 1929 paper. Figure 9.5(B) includes the velocity of astronomical objects at much greater distances than the observations presented by Hubble in his 1929 paper. Hubble and Humason (1931) used the redshift to observe more distant objects.

The linear relationship between radial velocity and distance shown in Figure 9.5 can be used to estimate the rate of expansion of the universe. For example, the dashed line in Figure 9.5(A) is a linear regression fit to the data, while the solid line is based on the estimated slope H_0 of 500 km/s/Mpc, where 1 Mpc is a megaparsec. By contrast, the dotted line

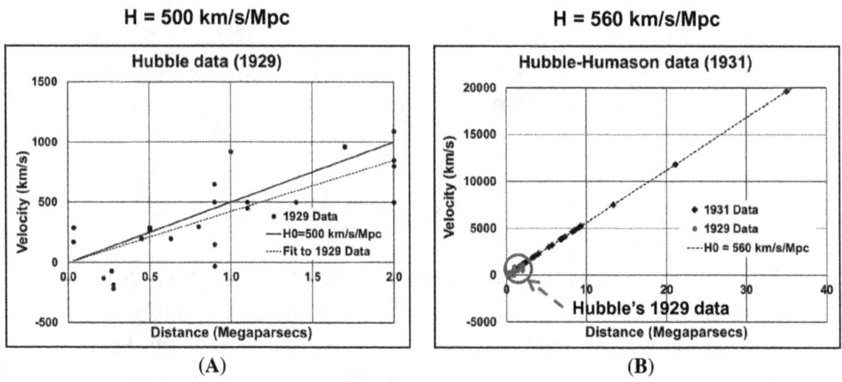

(A) (B)

Figure 9.5. Hubble Diagrams.

in Figure 9.5(B) is based on the estimated slope H_0 of 560 km/s/Mpc. The slope H_0 of the line drawn through the points is the rate of expansion of the universe and is now known as the Hubble constant or Hubble's parameter. The modern estimate of the rate of expansion H_0 is much lower, as we explain below.

9.5 Lemaître's Cosmogony

By 1931, Hubble's velocity–distance diagrams provided evidence for an expanding universe. The expansion of the universe is detected by observing light from the past. Light from light-emitting astronomical objects such as stars or galaxies show us what the astronomical objects looked like in the past when the light was emitted. The speed of light in vacuum is approximately 300,000 km/s or about 186,000 mi/s. This large but finite speed takes time to reach Earth in our vast universe. For example, the Andromeda Galaxy is a large spiral galaxy relatively close to the Milky Way Galaxy. It is about 2.5 million light years away. Light from a distant star or galaxy takes longer to reach Earth than light from a nearby star or galaxy.

The idea that the expanding universe began in a fireball was first proposed by Lemaître in 1931. He presented a model of the origin of the universe, or cosmogony, that used a 'primeval atom' to initiate an explosive expansion of the universe known as a 'primeval fireball'. Lemaître's model is known as the hypothesis of the primeval atom.

According to Lemaître, the expanding universe began from a primeval atom:

> **Q9.9**. "The radius of space began at zero; the first stages of the expansion consisted of a rapid expansion determined by the mass of the initial atom, almost equal to the present mass of the universe." (Lemaître, 1933, p. 52)

The expansion of the universe proceeded in three phases:

> **Q9.10**. "A first period of rapid expansion in which the atom-universe was broken into atomic stars, a period of slowing-down, followed by a third period of accelerated expansion. It is doubtless in this third period that we find ourselves today, and the acceleration of space which

> followed the period of slow expansion could well be responsible for the separation of stars into extra-galactic nebulae." (Lemaître, 1933, p. 52)

Lemaître's hypothesis of the primeval atom was the precursor of the Big Bang theory.

# # #

Albert Einstein published the general theory of relativity in 1916 and applied it to cosmology in 1917. He originally embraced a static model of the universe because there was a lack of evidence supporting either an expanding or contracting universe. Alexander Friedmann and Georges Lemaître showed that solutions of the field equations of the general theory of relativity could model a static universe, a contracting universe, or an expanding universe. By 1931, astronomers like Vesto Slipher and Edwin Hubble had provided enough evidence of an expanding universe to convince Einstein to abandon the static universe and accept the idea that the universe was expanding (Nussbaumer, 2014). Does an expanding universe imply that the universe had a beginning? If the universe had a beginning, did space and time also have a beginning? These ideas are explored in more detail in the following chapters.

Endnote

1. Sources on cosmology include Einstein (1917), Pais (1982), Peebles (1993), Weinberg (1993, 2008), Kragh (1996), Liddle (1999), Singh (2004), Kisslinger (2017), Perlov and Vilenkin (2017), Bennett *et al.* (2018), Fraknoi *et al.* (2018), Odenwald (2019), Teerikorpi *et al.* (2019), Malkan and Zuckerman (2020), and Hartle (2021).

Chapter 10

Evidence for the Expanding Universe

German astronomer Heinrich Wilhelm Olbers (1758–1840) wondered why the sky was dark at night.[1] At the time, the universe was considered infinite, eternal, and unchanging. If the universe was infinitely old, infinitely large, and contained an infinite number of stars, every line of sight from Earth would end at a star, as illustrated in Figure 10.1. In this case, it would be reasonable to believe that the night sky should be completely illuminated, yet the night sky was dark with a scattering of spots of light. The seeming contradiction between what people believed and what they observed was called Olbers' paradox or the dark night sky paradox.

Olbers' paradox could be resolved by disputing the assumptions. For example, the universe was not infinitely old, so light from all stars has not yet reached Earth. Alternatively, the universe does not contain an infinite number of uniformly distributed stars, so there are dark areas in the night sky. A more modern explanation of Olbers' paradox is that the universe is expanding, the distribution of light sources is not uniform, and light from distant light sources has not had enough time to reach Earth.

Scientists believe that we live in an expanding universe that exists in the aftermath of a cosmic explosive event. We examine more evidence for an expanding universe in this chapter[2] to justify why it is important to seriously consider proposals such as Lemaître's hypothesis of a primeval atom or, more recently, the Big Bang. We begin by discussing how astronomical distances are measured.

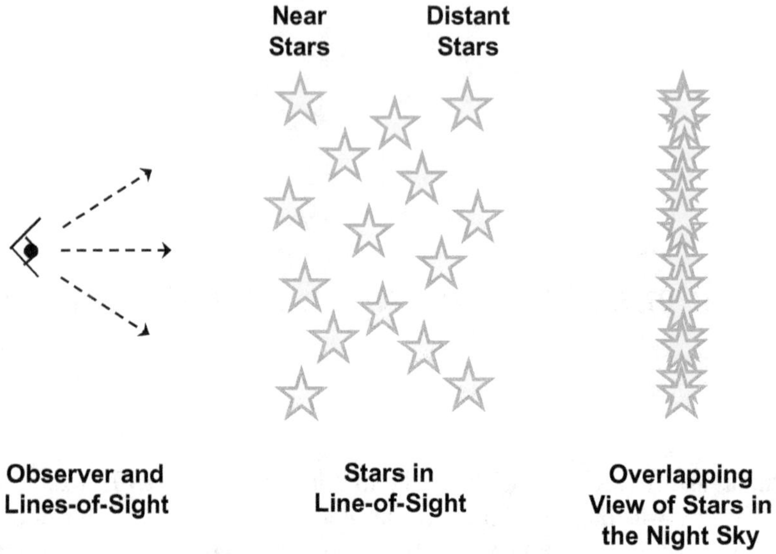

Near Stars

Distant Stars

Observer and Lines-of-Sight

Stars in Line-of-Sight

Overlapping View of Stars in the Night Sky

Figure 10.1. Schematic of Olbers' Paradox.

10.1 Cosmic Distance Ladder

Hubble's law is based on Doppler measurements and distances to galaxies. It shows that galaxies farther from Earth are moving away from us at a greater velocity than galaxies closer to Earth. Measurements show a linear relationship between the velocity of galaxies relative to Earth and their distance from Earth.

More than one method is needed to measure astronomical distances because the accuracy of the method depends on the distance to an astronomical object. Methods for measuring intermediate to extreme distances depend on methods used to measure distances to nearby astronomical objects. Methods that overlap can be used to calibrate when the range of applicability of the methods overlaps.

Astronomical distance measurement methods form a cosmic distance ladder where higher rungs on the ladder are used to measure greater distances than lower rungs. Methods that contribute to the cosmic distance ladder are discussed here, beginning with a method capable of measuring distances to nearby stars.

Stellar parallax

Triangulation is a method for measuring the distance to a point on a triangle by measuring angles to the point from the two points at opposite ends of a baseline with known length. Stellar parallax is a triangulation method that measures the distance to a nearby object such as a star using two different lines of sight, as illustrated in Figure 10.2. A triangle is formed with the nearby object at one apex and observations from two points of the Earth's orbit at the remaining two apexes. The position of the nearby object appears to shift relative to a field of distant stars when viewed from two different lines of sight. The most accurate measurement of distance to the nearby object is achieved by maximizing the length of the baseline. The longest baseline is formed when Earth-bound observations are made from opposite sides of the Earth's orbit around the Sun. The distance to the nearby object is calculated using the baseline length between observations and angles to the lines of sight shown in Figure 10.2.

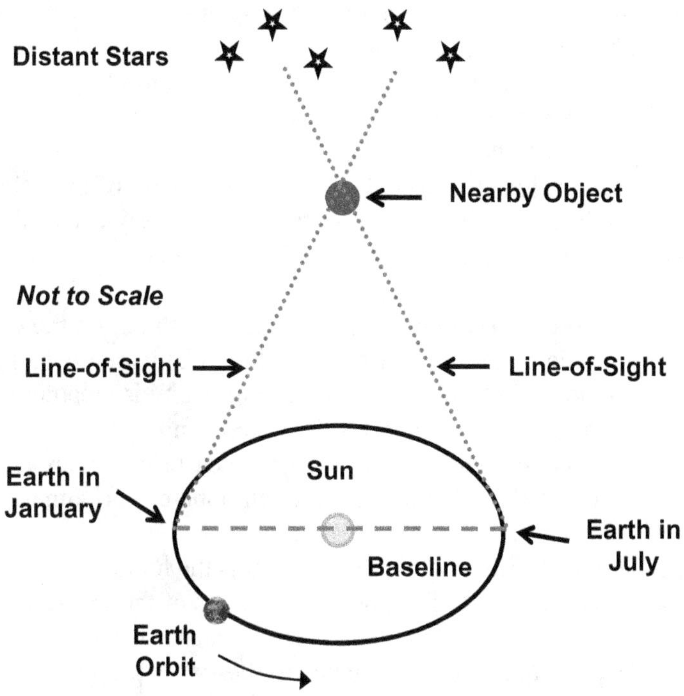

Figure 10.2. Stellar Parallax.

The parallax method can estimate distances up to approximately 10,000 light years. Larger distance measurements are needed to measure the distance to astronomical objects that have been observed emitting light from distances stretching over billions of light years. Light emitted by these objects began its journey to the Earth billions of years ago. We get a glimpse of the universe as it appeared near the time of the Big Bang billions of years ago when we observe distant astronomical objects. Stellar parallax does not work at these distances, so we need new methods.

Cepheid variable stars

A method that provides distance measurements to distances up to 100 million light years is based on a discovery by Henrietta Swan Leavitt (1868–1921) at Harvard Observatory (Leavitt and Pickering, 1912; Leavitt, 1908). She observed the periodic dimming and brightening of visible light originating from some stars in the Small Magellanic Cloud. These pulsating stars are known as Cepheid variables.

The length of time, or period, of the pulsation is the length of time it takes for the pulsation to brighten, dim, and brighten again. The period is related to the luminosity of a Cepheid variable, where luminosity is the total energy radiated each second by a star. The period of pulsation of a Cepheid variable is its 'fingerprint' or 'signature'. Leavitt established a relationship between luminosity of a Cepheid variable and the period of pulsation.

Astronomers have developed a procedure for estimating the distance to Cepheid variables using Leavitt's period–luminosity relationship. First, measure the period of a Cepheid variable. Luminosity is then determined using the period–luminosity relationship. Once luminosity is known, the distance to the Cepheid variable is found by measuring apparent brightness and then using the relationship between apparent brightness, luminosity, and distance.

The apparent brightness of a spherical star is the luminosity of the star divided by the surface area of a sphere. The radius of the sphere begins at the center of the star and is equal to the observer's distance from the star (Figure 10.3). The square of the distance to the star is proportional to the ratio of luminosity to apparent brightness. The distance between the star and the observer increases as the apparent brightness of the star decreases.

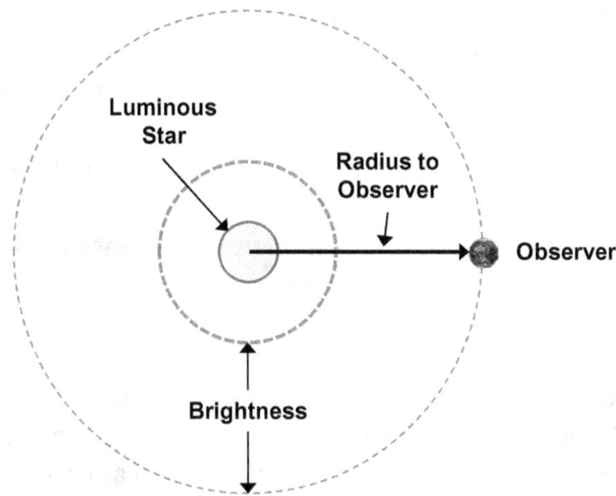

Figure 10.3. Stellar Brightness Decreases with Distance from the Observer.

Once we know the distance to a Cepheid variable, we can estimate the distance to neighboring light sources. Edwin Hubble used Cepheid variables and Doppler-shifted stellar spectra to measure distances to galaxies.

Supernovae

Cepheid variables are not well suited for measuring the distances of the most remote objects we can detect. Another astronomical discovery was needed, and it came in the form of supernovae. A supernova is a star that ends its life in an explosion that can be seen over billions of light years.

A supernova is a Type II supernova if it displays hydrogen lines in its spectrum, otherwise it is a Type I supernova. The term 'supernova' refers to the explosive destruction of a star. As massive stars age, available fuel is consumed by the fusion process, which is a nuclear process that combines small atomic nuclei into larger atomic nuclei. For example, under sufficient temperature and pressure, hydrogen nuclei can fuse to form a helium nucleus. Eventually, the fusion process reaches a stage when the repulsive pressure associated with the release of energy by nuclear fusion is overcome by attractive gravitational forces associated with the mass of the star. The subsequent collapse of the stellar core results in a Type II supernova.

Type II supernovae typically have masses that exceed the mass of the Sun by many times. A Type I supernova can occur with less massive stars. The Type Ia supernova is an important example of the cosmic distance ladder. Consider a white dwarf star with an inert carbon–oxygen core orbiting a companion star in a binary star system. If the gravitational attraction of the white dwarf star is enough to attract matter from its companion star, it can accumulate enough matter to increase its core temperature to the point where carbon fusion is initiated. The fusion process leads to an explosive nuclear reaction. The resulting Type Ia supernova has a characteristic light curve that displays a time-varying luminosity.

The Type Ia supernova is an example of a standard candle. In the context of astronomy, a standard candle is a light source with known luminosity. By comparing a nearby standard candle such as a Cepheid variable star with the light curves of distant standard candles such as a Type Ia supernova, we can calibrate the Type Ia supernova as a standard candle. A calibrated Type Ia supernova can be used to measure distances to neighboring light sources at distances that exceed distances measured by parallax methods or Cepheid variable stars.

Cosmic distance ladder

The methods for measuring astronomical distances described above illustrate methods that contribute to the cosmic distance ladder. The cosmic distance ladder is also known as the cosmological distance scale or the cosmic distance chain. The Stellar Parallax rung in Figure 10.4 uses methods that can measure astronomical distances up to 10,000 light years, while the Cepheid variable stars rung measures astronomical distances between 1,000 light years and 100 million light years. The rung on the ladder named Distant Standards includes standard candles such as Type Ia supernovae calibrated with Cepheid variable stars. Distant standard candles can measure astronomical distances ranging from 10 million light years to billions of light years.

10.2 Expansion of the Universe

Astronomers continued plotting distances from Earth to distant galaxies and their Doppler shifts. They found that light from distant galaxies was redshifted and supported Hubble's law, that is, the relationship between

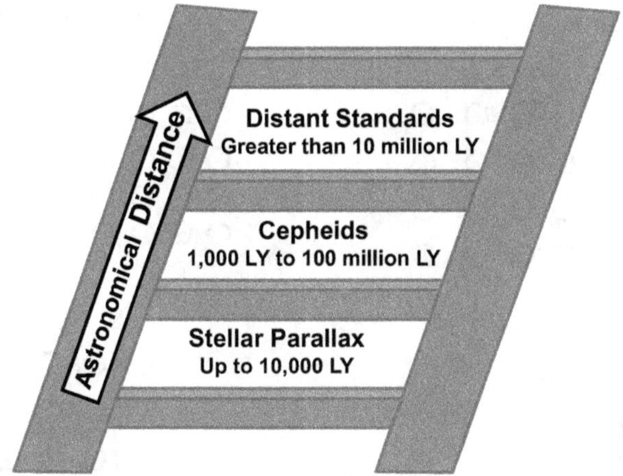

Figure 10.4. Illustration of a Cosmic Distance Ladder (LY = Light Years).

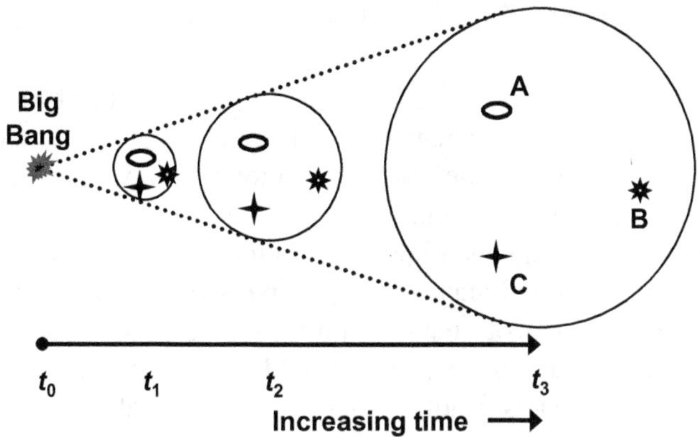

Figure 10.5. Illustration of the Expanding Universe.

recessional velocity of the galaxy and distance to Earth is linear. The slope of the straight line H_0 on a Hubble diagram is the Hubble constant, which quantifies the expansion rate of the universe. Galaxies throughout the universe are receding from Earth. How can the apparent motion of galaxies be explained as a cosmological phenomenon?

Astronomical objects are receding from one another as the space between them expands (Figure 10.5). The three objects in the figure move

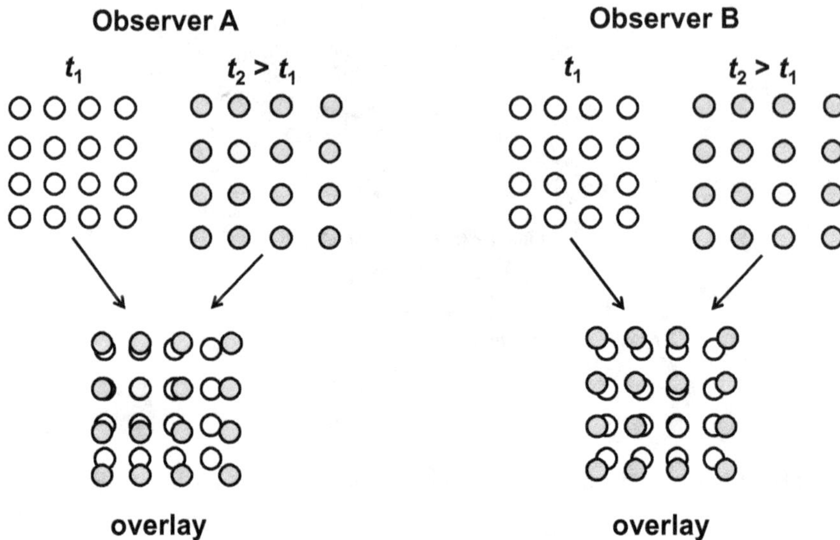

Figure 10.6. Each Observer Sees Expansion.

away from one another, and the separation between the objects increases as the surface expands. Light emanating from galaxies in an expanding universe is redshifted because the galaxies are moving away from one another.

The gravitational pull (curvature of spacetime) between massive objects in a galaxy helps the objects retain their relative positions within the galaxy. The motion of massive objects in regions of space with significantly greater mass density than the uniform mass density of a homogenous universe is predominantly influenced by the greater mass density in the region. Astrophysicist Andrew Liddle observed that:

> **Q10.1**. "if we go to large enough scales, in practical terms of megaparsecs, the Universe does become effectively homogenous and isotropic… It is on these large scales that the expansion of the Universe is felt, and on which the cosmological principle applies." (Liddle, 1999, p. 23)

Hubble postulated that the universe should look the same from any point in the universe if it is expanding from a uniform point. Figure 10.6 shows how Observer A and Observer B would see an expanding universe. The locations of the observers are indicated by the yellow circles. The set of circles in the top row on the left side of the figure shows what Observer A would see at an early time t_1 and a later time t_2. The right side of the figure shows what Observer B would see. The bottom row of circles shows an overlay of the circles in the top row. The overlay on the left-hand side is centered on Observer A, and the overlay on the right-hand side is centered on Observer B. Both observers see circles moving away from their location.

What is expanding?

Everywhere scientists look in the sky they see receding galaxies. It appears the universe is expanding, but what does that mean? The modern notion of an expanding universe was developed in the early 1900s. In his special theory of relativity, Albert Einstein (1905) replaced Isaac Newton's view (Newton, 1687) that time and space are absolute and independent. Einstein supported the idea that the three dimensions of space and one dimension of time were part of a four-dimensional spacetime continuum. The special theory of relativity applies only to non-accelerating bodies. Einstein published the general theory of relativity a decade later (Einstein, 1916) that applied to accelerating bodies, such as a massive object moving in a gravitational field. According to the general theory of relativity, the universe expands as the spacetime continuum expands.

Does the speed of recession have a limit?

Hubble's law says that two objects seem to separate from each other at a speed that exceeds the speed of light when the objects are far enough apart. It would appear that this violates one of the postulates made by Einstein in his special theory of relativity:

> **Q10.2**. "Light is always propagated in empty space with a definite velocity c which is independent of the state of motion of the emitting body." (Einstein, 1905, p. 38)

Einstein argued that the speed of light is constant in empty space, and that matter cannot travel faster than the speed of light. Hubble's law and Einstein's postulate of the constancy of the speed of light are consistent with each other if we recognize that Hubble's law refers to the rate of expansion of spacetime, while Einstein's postulate refers to the motion of an object within spacetime.

10.3 Cosmic Microwave Background Radiation

Advocates of an expanding universe suggest that the expanding universe has resulted from an explosive cosmic event. Hubble's law and an explosive origin of the universe suggest that the expanding universe is evolving away from its origin (Figure 10.5). If we trace the present position of galaxies back in time, can we find the point of origin of the universe? When and where do galaxies converge? Is it possible to find evidence of the explosion?

Evidence of a Big Bang was discovered in 1964 by Americans Arno A. Penzias (1933–) and Robert W. Wilson (1936–). Their employer, Bell Laboratories, wanted to determine if microwave radiation could interfere with satellite communications. Penzias and Wilson chose a sensitive instrument known as a horn reflector or microwave antenna to observe light with wavelengths ranging from approximately 7 to 20 cm (about 3 to 8 in).

Radiation from space covers a relatively broad range of spectral frequencies that can be loosely characterized as noise. Some of this noise could be inherent to an electrical device. For example, it could arise as interference from Earth-based radiating systems such as television or radio transmitters, or it could originate in the cosmos. Prior to 1964, no one had successfully isolated cosmic microwave frequencies in a study of radiation streaming to Earth from outer space. Penzias and Wilson needed to remove all microwave radiation that did not come from space.

Penzias and Wilson expended great effort trying to eliminate unwanted radiation sources. When they made their observations, they detected a signal — a hiss — coming from everywhere in the sky and in all seasons. The energy of the radiation they observed was equivalent to an absolute temperature of about 3.5 K or about –453°F. What was the radiation?

The radiation observed by Penzias and Wilson is now considered a relic of the primordial explosion that initiated the expanding universe. It is called the cosmic microwave background radiation (CMB). The discovery of the CMB helped make cosmology an experimental science.

A spacecraft designed to improve our view of the early universe (NASA-COBE, 2009) was launched in November 1989. The Cosmic Background Explorer (COBE) spacecraft operated from 1989 to 1993. The four years of CMB temperature observations were used to create the map shown in Figure 10.7. The CMB map displayed anisotropy in temperature for temperature fluctuations in the range of $\pm 100 \times 10^{-6}$ K.

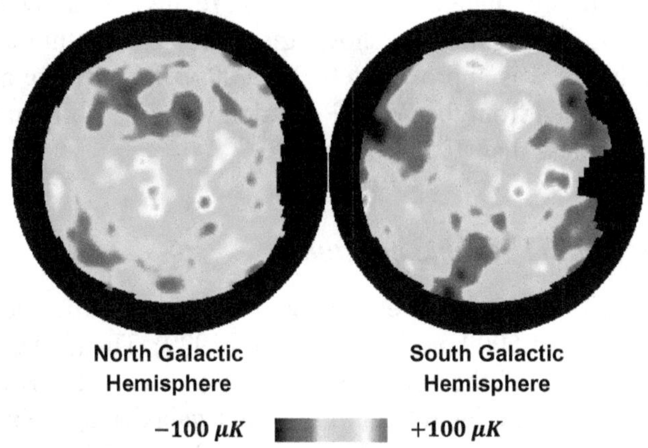

Figure 10.7. COBE-DMR Map of CMB Anisotropy (NASA-COBE, 2009).

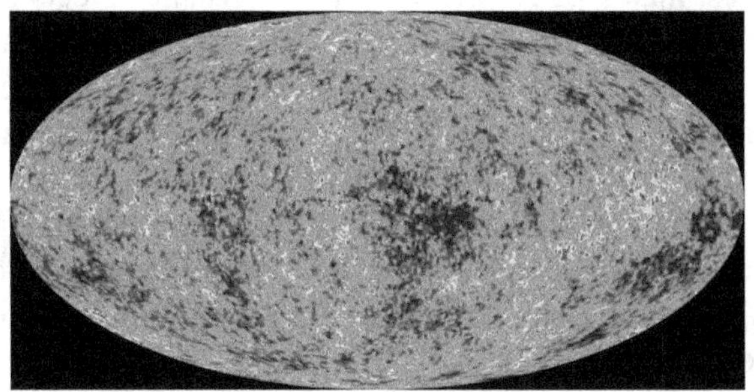

Figure 10.8. WMAP of CMB Temperature Anisotropy (NASA-WMAP, 2010).

The CMB temperature fluctuations presented in the figure show temperature differences between the 2.73 K average temperature of the radiation field and the observed value. Temperature fluctuations range from approximately -100×10^{-6} K in blue to $+100 \times 10^{-6}$ K in red. The data were measured by a differential microwave radiometer. The map shows the CMB anisotropy when the universe was approximately 380,000 years old.

The Wilkinson Microwave Anisotropy Probe (WMAP) provided more accurate measurements of the CMB anisotropy observed by COBE. The WMAP spacecraft was launched in 2001 (NASA-WMAP, 2010, 2011) and observed five different microwave frequencies in the full sky. The anisotropy map in Figure 10.8 is based on 9 years of data and shows a temperature range of $\pm 200 \times 10^{-6}$ K relative to an average temperature of 2.73 K. Figure 10.8 shows the CMB temperature anisotropy when the universe was approximately 375,000 years old. The age of the universe determined from the WMAP mission was 13.77 billion years old with an error of less than 1%.

Discrepancy in the Hubble constant

The earliest estimate of the expansion rate of the universe H_0 was Hubble's estimate of 500 km/s/Mpc (Hubble, 1929). The next estimate of the expansion rate of the universe H_0 was provided by Hubble and Humason (1931). They included more distant galaxies and obtained the expansion rate H_0 to be 560 km/s/Mpc. The earliest estimates of H_0 were significantly greater than modern estimates of H_0 provided by Riess *et al.* (2016) and the Planck collaboration (2020).

Modern estimates of H_0 included more distant objects and found a significant decrease in H_0. Riess *et al.* (2016) used the Hubble Space Telescope to observe Cepheid variables and estimated H_0 to be 73.4 km/s/Mpc with uncertainty ±1.74 km/s/Mpc. The Planck collaboration (2020) used CMB data from the European Space Agency's Planck satellite to estimate H_0 to be 67.4 km/s/Mpc with uncertainty ±0.5 km/s/Mpc. These two estimates are based on independent methods and give comparable values of H_0, but the values of H_0 and associated uncertainties do not overlap. The discrepancy in H_0 values is an example of an anomaly in normal science.

# # #

The interpretation of Hubble's law seemed to favor the idea of a universe that was expanding from a much more compact accumulation of mass and energy. Observations of recessional velocity–distance relations and the CMB further supported the idea of universal expansion. The hypothesis of the primeval atom by Lemaître in 1931 was the first attempt to provide a mechanism for initiating an explosive expansion of the universe following a moment of creation. Today, cosmology is based on the idea that the universe began with an explosive release of mass and energy on the subatomic scale, that is, the scale of objects smaller than an atom. In the next chapter, we review physics of objects on the subatomic scale that could have contributed to a cosmic explosive event.

Endnotes

1. Sources on Olbers paradox include Liddle (1999), Britannica (2018), and Odenwald (2019).
2. Sources on the expanding universe include Peebles (1993), Weinberg (1993, 2008), Liddle (1999), Singh (2004), Kisslinger (2017), Perlov and Vilenkin (2017), Bennett *et al.* (2018), Fraknoi *et al.* (2018), Odenwald (2019), Teerikorpi *et al.* (2019), Malkan and Zuckerman (2020), and Hartle (2021).

Chapter 11

Physics of the Subatomic Scale

Three classes of universes are possible: open universe models, flat universe models, and closed universe models. We saw in Figure 8.6 that the geometry of a flat universe is Euclidean, the geometry of an open universe is hyperbolic non-Euclidean, and the geometry of a closed universe is elliptic non-Euclidean. Illustrations of open universe and closed universe models are presented in Figure 11.1.

Open universe models assume the universe began as an event in spacetime. Spacetime began with the release of all the mass–energy of the universe in the Big Bang. Following the initial event, the spacetime domain containing the mass–energy of the universe expands in the aftermath of the Big Bang. The expansion of spacetime will continue indefinitely following the initial event in open universe models.

By contrast, the expansion of the spacetime domain will eventually stop in closed universe models. Spacetime then begins to contract until all of the mass–energy of the universe is once again compressed to a single spacetime point. Another explosive release of mass–energy from the single point may occur, but the mechanism is still unknown. If an explosive release of mass–energy does occur, spacetime once again starts to expand. In oscillating universe models, the expansion and contraction phases periodically repeat.

Spacetime geometry of possible universes can be considered on two scales: local and global. The scales range from subatomic size to the size of the universe. Global spacetime geometry describes the curvature of spacetime of the entire universe, while local spacetime geometry describes the curvature of spacetime in regions that are defined on a

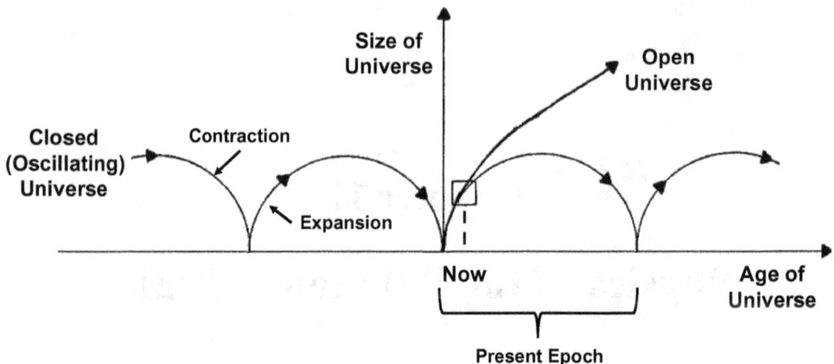

Figure 11.1. Open Universe and Closed Universe Models.

smaller scale. To understand the origin and evolution of the universe, we must consider the physics of mass and energy on the subatomic scale. This requires discussion of another great advance in 20th century physics: quantum theory.

11.1 Transition from Classical Physics to Quantum Physics

Our focus in Chapters 4 through 10 has been classical physics and the macroscopic scale. Scientists before 1900 were able to explain the observations of almost all known experimental measurements using classical physics. Observations from a few experiments remained anomalous. An eventual understanding of the anomalous experiments required reconsideration of many widely accepted concepts. The beginning of the 20th century saw the emergence of a new physics.

An example of an anomalous observation we have already discussed is the constancy of the speed of light in vacuum. Albert Einstein (1879–1955) provided an acceptable explanation when he introduced the theory of special relativity, but he had to reject Isaac Newton's (1643–1727) concepts of absolute and independent space and time.

The radiation emitted by heated black bodies is another example of an anomalous observation. It was explained by the rejection of a previously accepted concept that energy was continuous. The black body experiment and the physical theory emerging from it are now considered.

Black body radiation

The cosmic microwave background radiation discussed in Chapter 10 appears to be black body radiation. A black body is an insulated, hollow enclosure, or cavity, with a small hole in one side, as illustrated in Figure 11.2. Heat or radiation cannot enter the cavity through its exterior walls. The interior walls of the cavity are designed to absorb all incident radiation frequencies in the cavity. The walls radiate energy that is characteristic of the temperature of the cavity. Black body radiation is studied by measuring the wavelength of the radiation which escapes from the small hole in the side of the cavity.

Energy density of the emitted radiation is plotted against wavelength in Figure 11.3. The dashed curve labeled 'UV Catastrophe' in the figure

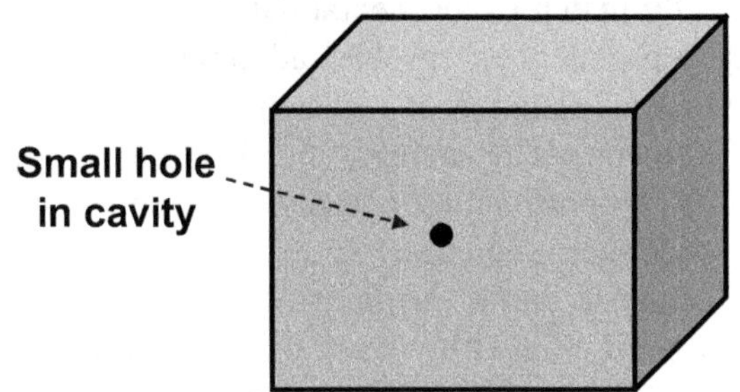

Figure 11.2. A Black Body.

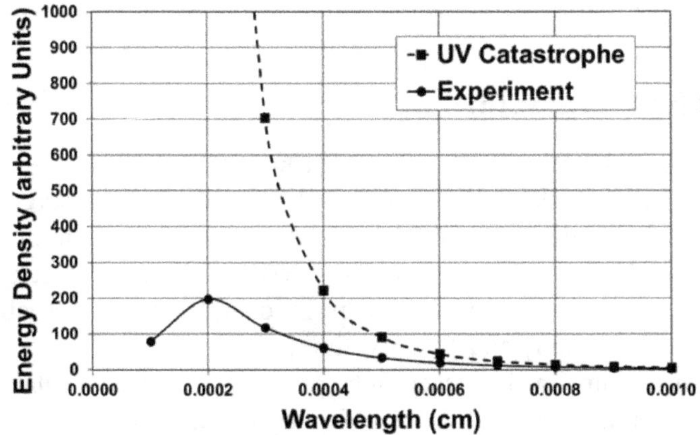

Figure 11.3. Black Body Radiation.

shows the calculation of the energy density of black body radiation based on classical physics. The classical theory, that is, the theory based on classical physics, assumes that energy is a continuous function of wavelength. The classical theory predicts that the energy density of black body radiation emitted at very short wavelengths is very large. Experiments, however, show that the energy density becomes small as the wavelength approaches zero. The discrepancy between classical theory and experimental observations at low wavelengths is called the 'ultraviolet catastrophe'. Does this discrepancy imply that something is wrong with the classical theory?

German physicist Max Planck (1858–1947) provided an answer[1] to this question in 1900. The curve labeled 'Experiment' in Figure 11.3 shows that Planck's theory matched all of the experimental data, including observations of emitted radiation at very short wavelengths. Planck was able to theoretically match experimental data by assuming that radiation exists as a discrete packet of energy. He had to replace the assumption that energy is a continuous function of wavelength. Planck's discrete energy packets are called quanta. The energy of the quantum is proportional to its frequency, and the proportionality constant is a fundamental constant of nature now called Planck's constant.

Planck's hypothesis that a quantum of light is a discrete packet of energy has been substantiated by many experimental tests and applications. For example, we saw in Chapter 8 that Einstein treated light as a discrete packet, or quantum, of energy in his 1905 paper entitled *On a Heuristic Point of View Concerning the Production and Transformation of Light* (Einstein, 1905c).

11.2 The Uncertainty Principle

German physicist Werner Heisenberg (1901–1976) proposed the uncertainty principle[2] in 1927. He showed that simultaneous measurements of certain pairs of physical variables must satisfy a mathematical inequality. The pairs are known as complementary variables and can characterize the state of a particle. Examples of pairs of complementary variables are {position, momentum} and {energy, time}.

The inequality in the uncertainty principle is called an uncertainty relation. It limits how much information we can get when we simultaneously measure values of pairs of complementary variables. As an illustration, an uncertainty relation for position and momentum says that we cannot simultaneously measure values of both position and momentum with 100% certainty. If we design an experiment to exactly measure the position of an object, then the position–momentum uncertainty relation says that the experiment cannot provide any information about its momentum. Similarly, if the experiment is designed to exactly measure the momentum of an object, then the position–momentum uncertainty relation says that the experiment cannot provide any information about the object's position.

Heisenberg suggested that nature imposes a limit to our knowledge. This limitation is not something we can reduce by improving our measuring techniques. According to Heisenberg:

> **Q11.1**. "...in classical physics it has always been assumed either that this interaction [between observer and observed object] is negligibly small, or else that its effect can be eliminated from the result by calculations based on 'control' experiments. This assumption is not permissible in atomic physics; the interaction between observer and object causes uncontrollable and large changes in the system being observed, because of the discontinuous changes characteristic of atomic processes." (Sambursky, 1975, p. 518)

French physicist and aristocrat Louis Victor de Broglie (1892–1987) observed that Heisenberg's uncertainty principle makes it

> **Q11.2**. "...impossible to attribute simultaneously to a body a well-defined motion and a well-determined place in space and time." (de Broglie, 1966, p. 122)

If we cannot simultaneously specify the location and momentum of an object with certainty, then the determinism of classical physics fails.

The energy–time uncertainty relation

The energy–time uncertainty relation says that there is a limit to how well we can simultaneously measure the duration and energy of a quantum process. Another way to understand the energy–time uncertainty relation is to recognize that a large amount of energy can appear for a short duration of time specified by the energy–time uncertainty relation. Particles that appear for the short duration allowed by the energy–time uncertainty relation are called virtual particles.

The energy–time uncertainty relation is needed to understand the vacuum process shown in Figure 11.4. The vacuum process is displayed as a Feynman diagram, which was introduced in Chapter 7 (Figures 7.2 through 7.4). As a reminder, we show that the Feynman diagram is a portion of a spacetime diagram. Particles are represented by arrows and an interaction occurs at a vertex. The space and time axes are shown as dotted lines in Figures 11.4(B) and 11.4(C) because the axes may or may not appear in a Feynman diagram. If space and time axes are not shown, they are implied.

The vacuum process in Figure 11.4(C) shows the creation, interaction, and annihilation of a proton (p), an antineutron ($\overline{n}$), and a pion (π^-). Antimatter particles have the same mass as matter particles but differ in some properties. For example, the neutron and antineutron are electrically neutral but differ in baryon number. The neutron has baryon number 1, and the baryon number of the antineutron is –1. The meaning of baryon is discussed in more detail in Chapter 12. Another example of a matter–antimatter pair is the negatively charged electron e^- and positively charged positron e^+.

No particles exist in Figure 11.4(C) before or after the vacuum process. The energy of the particles in the vacuum process exists for the short duration of time allowed by the energy–time uncertainty relation. This appears to be true even if the conservation of energy is violated for the duration allowed by the energy–time uncertainty relation. The

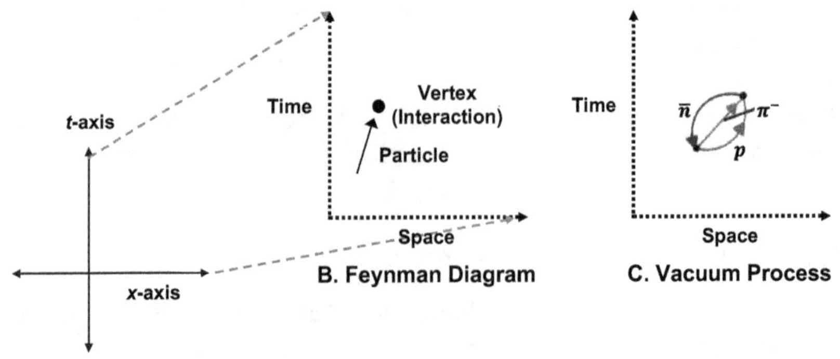

Figure 11.4. Feynman Diagram for a Vacuum Process.

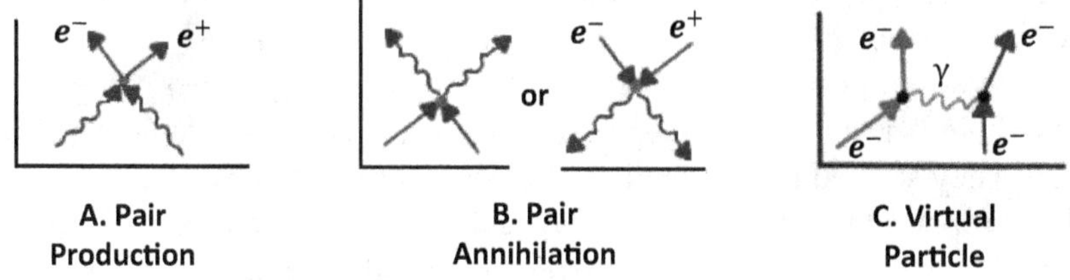

Figure 11.5. Examples of Particle Interactions.

short-duration change in energy is an example of a quantum fluctuation. We show in Chapter 12 that quantum fluctuations are a possible natural mechanism for initiating the Big Bang.

Some additional particle interaction processes are shown in Figure 11.5. The Feynman diagrams in the figure depict pair production, pair annihilation, and the formation of a virtual particle when two electrons with the same charge repel each other. The wavy line in each of the figures represents a photon or particle of light. The symbols e^- and e^+ denote an electron and its antiparticle, the positron, respectively. Matter and antimatter are produced in the pair production process as a matter–antimatter pair when sufficient electromagnetic energy is available from the photons. When a matter–antimatter pair interacts, they annihilate each other and transform into electromagnetic radiation.

Figure 11.5(C) shows the electric repulsion between two electrons. The electrons have the same electrical charge, and we know from experiments that the electrons should repel each other. The repulsion occurs when the electrons exchange a photon labeled by the lowercase Greek letter gamma (γ). The photon is a virtual particle because it only exists for the duration allowed by the energy–time uncertainty relation. The photon is considered a carrier of the repulsive force. The limiting value of the speed of light in vacuum can be honored when we treat interactions as the exchange of virtual particles, such as a photon.

11.3 Complementarity and Wave–Particle Duality

In our discussion thus far, we treated light like a wave in Chapter 4 and then we treated light like a particle in the photoelectric effect in Chapter 8. Louis Victor de Broglie observed in the 1920s that electromagnetic radiation could exhibit wave-like behavior and particle-like behavior.[3]

De Broglie hypothesized that all matter could exhibit wave-like behavior and the wavelength of matter was inversely proportional to wavelength. The wave-like properties of matter were confirmed by American physicists Clinton Davisson (1904–1991) and Lester Germer (1896–1971) in 1927. They studied the diffraction of electrons scattered by a nickel crystal and reported that electrons displayed diffraction patterns similar to the diffraction patterns of X-rays scattered by crystals (Davisson and Germer, 1927). One of the results of the Davisson–Germer experiment was the calculation of equivalent wavelengths of electron beams that were inversely proportional to electron momentum, which was a prediction of de Broglie's hypothesis.

Scientists realized that light and matter can exhibit wave-like or particle-like properties. Furthermore, experiments can be devised to show wave-like properties, such as diffraction, or particle-like properties, such as the photoelectric effect, but the experiments cannot simultaneously show both wave-like and particle-like properties. This phenomenon is called wave–particle duality.

The complementarity principle advocated by Danish physicist Niels Bohr (1885–1962) was presented as an explanation of wave–particle duality. Bohr's complementarity principle said that some objects have certain pairs of complementary properties which cannot be simultaneously observed or measured. For example, suppose we flip a two-sided coin which has heads on one side and tails on the other side. If we flip the coin, we can display either heads or tails but not both. Our simple experiment is unable to show both sides of the coin. Similarly, if we design and perform Experiment A to study the wave-like properties of light, we would observe light behaving like a wave. On the other hand, if we design and perform Experiment B to study the particle-like properties of light, we would observe light behaving like a particle. If we carefully analyze the experiments, we realize that our methods are unable to simultaneously measure the wave and particle properties of light.

Many people have disagreed about the nature of light (Table 11.1). Dutch scientist Christiaan Huygens explained the properties of light, such as reflection and refraction, using waves, while Newton argued that light behaved like particle-like corpuscles. Maxwell's equations supported the idea that light was a wave. On the other hand, Einstein received the Nobel Prize for using the idea that light was a quantum of energy to explain the photoelectric effect. The limitations in our experiments are not due to

Table 11.1. Milestones in the Understanding of Light.

Person	Period	Contribution
Christiaan Huygens	1629–1695	Wave theory
Isaac Newton	1643–1727	Corpuscular theory
Max Planck	1858–1947	Quantum theory
Albert Einstein	1879–1955	Photoelectric effect
Louis de Broglie	1892–1987	Wave–particle duality

any fault we can correct, but are natural limitations associated with Heisenberg's uncertainty principle.

Wave–particle duality may be thought of as complementary descriptions of reality. Each view is correct for a limited range of applications, while both views are necessary for a complete description. From his laboratory in Copenhagen, Bohr showed that classical physics could no longer be used to adequately describe the spectrum of the atom. He used the idea of quantized energy to correctly calculate the spectrum of hydrogen and advocated the complementarity principle.

Bohr believed complementarity was necessary to science. After a series of debates with Albert Einstein, who believed that quantum theory was incomplete, Bohr's view was adopted as the mainstream scientific interpretation of quantum mechanics and became known as the Copenhagen interpretation. It represents the state of a physical system as a mathematical function called the wave function. In this view, the wave function is a complete description of the subjective knowledge of the observer rather than the objective state of the observed system. New knowledge about the system changes the mathematical description of the relationship between physical systems. A change in our knowledge changes the wave function and can be interpreted as a collapse of the wave function.

11.4 Born's Probabilistic View

We seem to have replaced the determinism of classical physics with a seemingly paradoxical theory. How can we understand within a single theoretical framework the ideas presented above? According to German-born British physicist Max Born (1882–1970), the wave function of quantum theory should be interpreted in terms of probabilities.[4]

One way to understand Born's probabilistic interpretation, which is also referred to as Born's rule, is to perform a gedanken experiment designed to illustrate one interpretation of quantum theory. The purpose of the experiment is to trace the motion of a particle — an electron — in space and time.

We can trace the motion of an electron in one space dimension and one time dimension by designing a detection system consisting of two detectors and two clocks to measure the worldline of an electron. Detectors 1 and 2 tell us the location of the electron along the space axis. Clocks 1 and 2 mark the times when the electron passes through Detectors 1 and 2, respectively. The detectors have a finite width that prevents the exact measurement of the position of the electron, and the resolution of the clocks prevents the exact measurement of the time of passage of the electron.

Physical measurements always contain sources of uncertainty. Systematic error limits the accuracy of a measurement, which refers to how close the measurement is to a specific or 'true' value. Random error limits the precision of a measurement, which refers to our ability to reproduce measurements. To minimize the effects of uncertainty, the measurements are repeated many times and then averaged. The resulting average values are used to trace the electron's worldline within an uncertainty represented by the experimental error. All location and time measurements are uncertain to some extent.

Circled dots in Figure 11.6 show the average values of our measurements. The size of each circle represents the uncertainty: the larger the circle, the larger the uncertainty. A straight line marked 'Classical' is

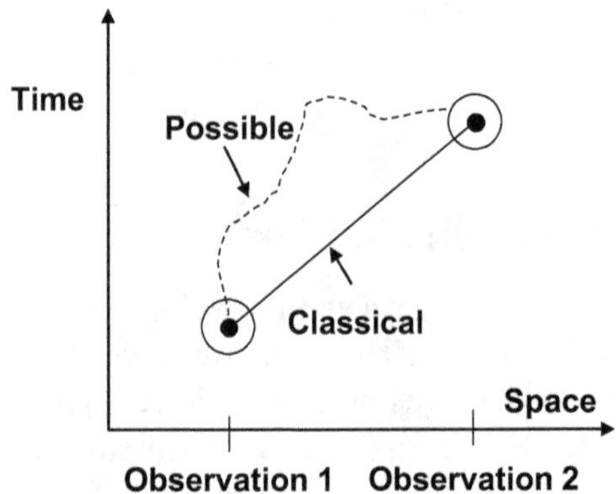

Figure 11.6. Possible Paths Between Observations.

drawn between the two measurements. This is the simplest worldline that can be drawn for the electron. The electron could have taken the more circuitous path marked 'Possible' and represented by the dotted line between the two observations. The most probable path is the classical path, even though alternative paths are possible. Paths other than the classical path play an important role in the modern scientific view of nature.

We have presented a procedure for measuring two points, or events, along the worldline. The path of the electron between the two events was unknown. Many paths between the events are possible. For each path, quantum theory lets us calculate the probability an electron would take that path. In many cases, the paths are not probabilistically equal, that is, one path is more likely, or more probable, than the other paths. The most probable path is the path determined by classical theory. Although classical theory tells us which path is most likely, it says nothing about alternative paths. Quantum theory, on the other hand, takes into consideration all possible paths and makes a statement about how likely each one is. Max Born is credited with introducing the historically significant idea that the calculated results of quantum theory should be interpreted in terms of probabilities.

Classical physics has successfully described a wide range of natural phenomena, particularly macroscopic phenomena. Quantum theory accounts for the success of classical physics by containing the classical theory as a special case of the more comprehensive quantum theory. Niels Bohr advocated for the Correspondence Principle, which said that quantum theoretical results become equivalent to the results of classical theory when quantum effects are negligible.

11.5 Interpretations of Quantum Theory

The most fundamental quantity of quantum theory is a function known by many names: the quantum state, the eigenfunction, the wave function, the 'psi' function Ψ, and so on. Born's work taught physicists to think of the quantum state within a probabilistic context. Even though physicists agree that the quantum state provides information about the likelihood of an event, they disagree on just how it should be interpreted. We consider three historically significant interpretations in this section: the Copenhagen interpretation, the statistical interpretation, and the Many-Worlds interpretation (MWI).

Scientists have used the formalism of quantum theory — the calculation procedure — without fully understanding all aspects of the formalism. An interpretation of a mathematically well-defined formalism does not have to be entirely correct before we can get useful information from it. Attempts have been made to highlight differences between interpretations. One notable attempt is Schroedinger's cat.

Schroedinger's cat

German physicist Erwin Schroedinger (1887–1961) proposed a gedanken experiment that highlights differences between interpretations. Suppose a live cat is placed in a closed chamber with a bottle of cyanide, a radioactive atom, and an automated hammer that will break the bottle when the radioactive atom decays. The scenario is illustrated in Figure 11.7. The decay of the atom is a chance occurrence; we know it will happen, but we do not know when.

The cat in Figure 11.7 and the observer in Figure 11.8 are macroscopic classical systems. The radioactive atom in Figure 11.7 is a

Figure 11.7. Initial Superposition of Quantum States.

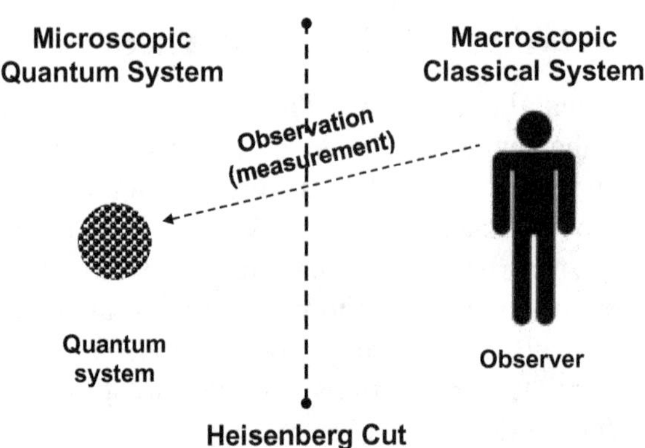

Figure 11.8. The Heisenberg Cut.

microscopic quantum system. We can hypothesize an interface between the classical system and the quantum system. The hypothetical interface is called the Heisenberg cut and is illustrated in Figure 11.8. The quantum system is represented by a quantum state, and the classical system is represented by a classical description.

When we look in the chamber, we do not know beforehand whether we will find a live cat or a dead cat. We can construct a quantum state S(live) describing the chamber with a live cat and an undecayed atom. We can also construct a quantum state S(dead) describing the dead cat and the decayed atom. The quantum state for the system as a whole S(system) is a superposition of the quantum states S(live) and S(dead). How should we interpret the quantum state S(system)? This question is the essence of Schroedinger's cat paradox.

According to the Copenhagen interpretation, S(system) completely describes everything we can know about Schroedinger's cat. If we look in the chamber when there is a 50% chance the atom has decayed, the quantum state S(system) will contain equal parts of S(live) and S(dead). Does that mean the cat is half dead and half alive? Copenhagen proponents answer no. They say the act of looking into the chamber forced the quantum state S(system) into one of the two quantum states S(live) or S(dead). This is known as collapse of the quantum state. Figure 11.9(A) shows the initial quantum state before an observation is made. Figures 11.9(B) and

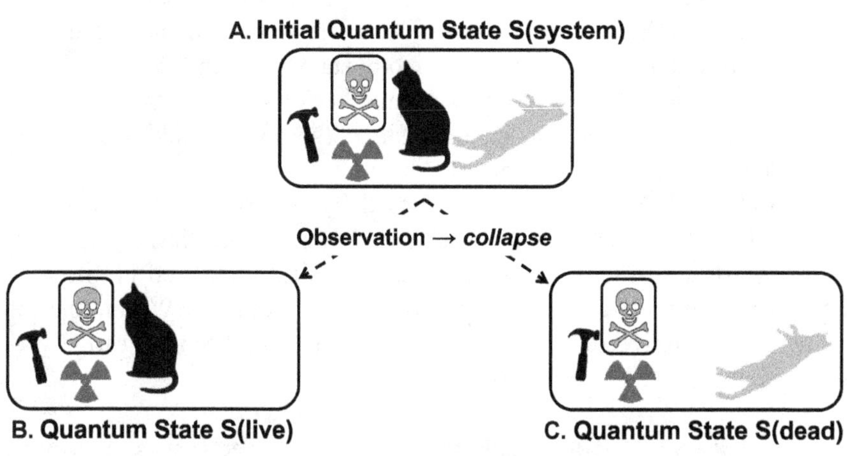

Figure 11.9. Copenhagen Interpretation.

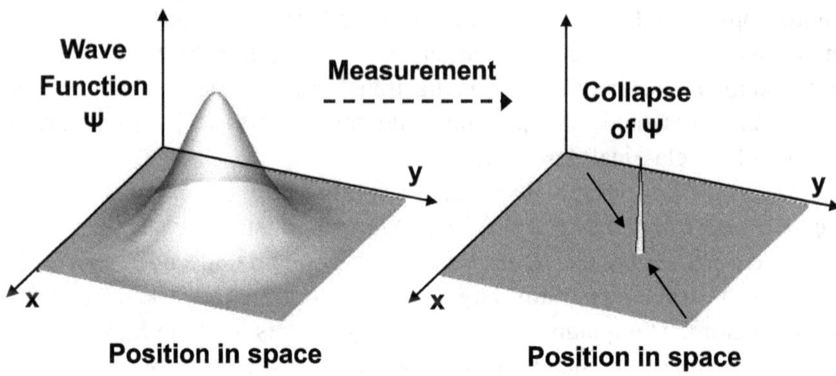

Figure 11.10. Collapse of a Quantum State.

11.9(C) show the collapse of the quantum state S(system) into one of the possible quantum states S(live) or S(dead).

Collapse of the quantum state is illustrated in Figure 11.10 for a quantum state represented as a spatially distributed wave function Ψ. The spatial distribution of the wave function is localized to a specific value when a measurement is made.

Proponents of the statistical interpretation take a different view. They envision a very large number of cats in the predicament devised by Schroedinger. The quantum state S(system) applies to each of the systems in this very large number of systems. If we look in all of the chambers when there is a 50% chance the atoms in each chamber have decayed, we will find half of the chambers contain dead cats and half contain live cats. In the statistical interpretation, the quantum state S(system) does not apply to a single system, but to a very large number of similarly prepared systems called an ensemble. The quantum state S(system) is making a statistical statement about an ensemble of systems.

Another interpretation[5] was proposed by American physicist Hugh Everett III (1930–1982) in 1957. Everett hypothesized that all possible outcomes of a quantum measurement are physically realized in different equally real 'worlds'. The interpretation is known as the MWI and the set of 'worlds' or universes is known as the multiverse. According to physicist Lev Vaidman,

> **Q11.3**. "a world is the totality of macroscopic objects: stars, cities, people, grains of sand, etc. in a definite classically described state." (Vaidman, 2021, Section 2.1)

The fundamental idea of MWI is that many 'worlds' coexist with our own. Every time a quantum experiment with different possible outcomes is performed, each outcome is realized in a different, newly created 'world'. MWI is described in more detail in our discussion of quantum gravity in Chapter 15.

Interpretations of the quantum state do not have all of the same features. For example, quantum state collapse in the Copenhagen interpretation is not a feature of either the statistical or MWI. Physicists have realized that they do not have to agree on the proper interpretation of the quantum state. Acceptance of one interpretation rather than another is not necessary to successfully apply quantum theory.

# # #

The processes diagrammed in Figures 11.4 and 11.5 have introduced six particles: the electron, positron, photon, proton, pion, and antineutron. These particles are six residents of the particle zoo. Finding an explanation for the number and type of particles that have been observed and the many ways they interact is a challenging, ongoing problem. The most widely accepted theory of particle physics is the Standard Model of Particle Physics. The particle zoo and the Standard Model of Particle Physics are discussed in Chapter 12. They contribute to our understanding of the science needed to provide a natural mechanism for initiating the Big Bang and, possibly, spacetime.

Endnotes

1. Sources on black body radiation include Brennan (1997), Cushing (1998), Kumar (2008), Kisslinger (2017), and Weinberg (2021).
2. Sources on the uncertainty principle include Heisenberg (1927), Brennan (1997), Cushing (1998), Kumar (2008), Weinberg (2015), and Teerikorpi *et al.* (2019).
3. Sources on wave–particle duality include Bacciagaluppi and Valentini (2009), Kumar (2008), and Weinberg (2015).
4. Sources on Born's probability interpretation include Born (1927), Cushing (1998), Kumar (2008), and Weinberg (2015).
5. Sources on the Many-Worlds interpretation include Byrne (2007), Barrett and Byrnes (2012), and Carroll (2019).

Chapter 12

Prelude to the Big Bang

Modern cosmology is based on the idea that the universe began with a Big Bang, that is, an explosive release of mass and energy. The hypothesis of the primeval atom by Lemaître in 1931 provided a mechanism for initiating an explosive expansion of the universe that began in a moment of creation. Observations of recessional velocity–distance relations and the cosmic microwave background further supported the idea of a Big Bang.

A natural initiation of the Big Bang depends on our knowledge of particle properties and the most widely accepted theory of the subatomic world, the Standard Model of Particle Physics.[1] We begin by identifying the four known fundamental interactions and their relationship to the classification of particles.

12.1 Fundamental Interactions and Particle Classification

Scattering and Feynman diagrams of two fundamental interactions were introduced in Chapter 7. So many different types of subatomic particles were discovered by scattering experiments that the collection of subatomic particles was dubbed the particle zoo. Scientists have learned that the occupants of the particle zoo can be classified by their interactions.

The four fundamental interactions are shown in Figure 12.1. They are the gravitational interaction, the weak interaction, the electromagnetic interaction, and the strong interaction. The electromagnetic and weak interactions are often combined and called the electroweak interaction.

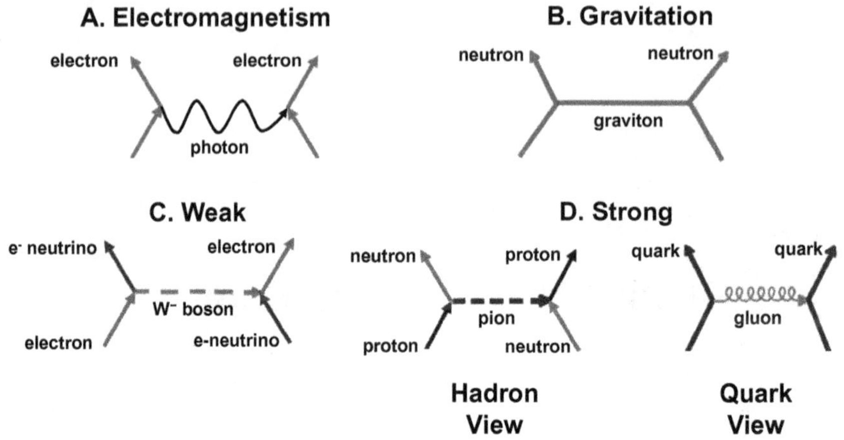

A. Electromagnetism
electron electron
photon

B. Gravitation
neutron neutron
graviton

C. Weak
e⁻ neutrino electron
W⁻ boson
electron e-neutrino

D. Strong
neutron proton
pion
proton neutron
Hadron View

quark quark
gluon
Quark View

Figure 12.1. Fundamental Interactions.

The strong and weak interactions are short-range interactions. Their influence dominates the behavior of interacting particles when the interacting particles are separated by distances comparable to the diameter of a nucleus or during the first few moments after the Big Bang. Electromagnetic and gravitational interactions are long-range forces. They exert an influence on particles separated by distances ranging from the diameter of an atom to the diameter of the universe.

The fundamental interactions are mediated by the exchange of a virtual boson between two fermions. A key difference between bosons and fermions is spin. Spin refers to the intrinsic angular momentum of a particle. It is a vector with spatial components in the *x*-, *y*-, and *z*-directions. The number 'm_s' in the term 'spin-m_s' is the spin quantum number. It refers to the projection of the spin vector along a particular axis that is commonly considered the *z*-direction axis. The spin quantum number tells us that the magnitude of the *z*-component of the spin angular momentum of a particle is m_s times a constant $\hbar$ (called h-bar). The constant $\hbar$ is $\hbar = \frac{h}{2\pi} = 1.546 \times 10^{-34}$ Joule · second where h is Planck's constant introduced in Chapter 11.

The spin quantum number m_s of a particle can be an integer $m_s = 0,1,2,\ldots$ or a half-integer $m_s = \frac{1}{2}, \frac{3}{2}, \ldots$. Fermions are particles with half-integer spin, and bosons are particles with integer spin. The bosons exchanged by fermions in the strong, electromagnetic, and weak interactions are spin-1 bosons, namely gluons, photons, and W^+, W^-, and Z^0

Table 12.1. Classification of Particles Based on Strong and Weak Interactions.

Level 1	Level 2	Level 3	Level 4
Photons γ			
Leptons (No Strong Interaction)	Neutrino ν		
	Electron e^- Positron e^+		
Hadrons (Strongly Interact)	Mesons (Weakly Interact)	Pion π	
		Kaon K	
		Eta η	
	Baryons (Distinct Particle–Antiparticle Pairs)	Nucleon	Proton p
			Neutron n
		Lambda Λ	
		Sigma Σ	

bosons. The graviton is a hypothetical spin-2 boson that mediates the gravitational interaction.

Table 12.1 presents a particle classification scheme based on strong and weak interactions. Level 1 shows two general categories of particles with mass. Massless particles of light (photons) are in a category of their own. Massive particles in Level 1 are called either leptons or hadrons. Hadrons can participate in the strong interaction, while leptons cannot. Examples of leptons are the neutrino, the electron, and the positron. Neutrinos do not interact electromagnetically. Electrons and positrons have electrical charge and interact electromagnetically.

The two hadron subgroups in Level 2 are mesons and baryons. Mesons can interact weakly, while baryons do not. Mesons, such as the pion, can be their own antiparticle, whereas baryons exist in distinct particle–antiparticle pairs. An example of a baryon pair is the proton and antiproton. Another common baryon is the neutron. The proton and neutron are also called nucleons because of their prominent role in atomic nuclei. The proton is a good illustration of the classification levels in Table 12.1. Starting with the proton in Level 4, we see that the proton is a nucleon, nucleons are baryons, and baryons are hadrons.

The gravitational and electromagnetic interactions are not as well suited for classifying particles as the strong and weak interactions. For example, all particles with mass can participate in gravitational interactions. By contrast, only electrically charged particles, such as electrons and protons, participate in electromagnetic interactions.

12.2 Atomism and Quarks

Historically, the treatment of matter as an aggregate of tiny, indivisible particles has been a useful approach. Prior to the beginning of the 20th century, atoms seemed to be the fundamental particles of nature. By the turn of the century, many observations were best explained by thinking of atoms as objects composed of smaller objects.

The atom illustrated in Figure 12.2 shows an atomic nucleus surrounded by an electron cloud. The nucleus contains most of the mass of the atom: positively charged protons and electrically neutral neutrons. The protons and neutrons are called nucleons because their behavior in the nucleus is controlled by the strong interaction. Electromagnetic repulsion between protons in the nucleus would disrupt the nucleus if the strong interaction was not present. Protons and neutrons are carriers of the strong force, and the strong interaction binds nucleons together. Both peaceful and military nuclear power depend on disrupting the strong force binding nucleons in the nucleus of an atom.

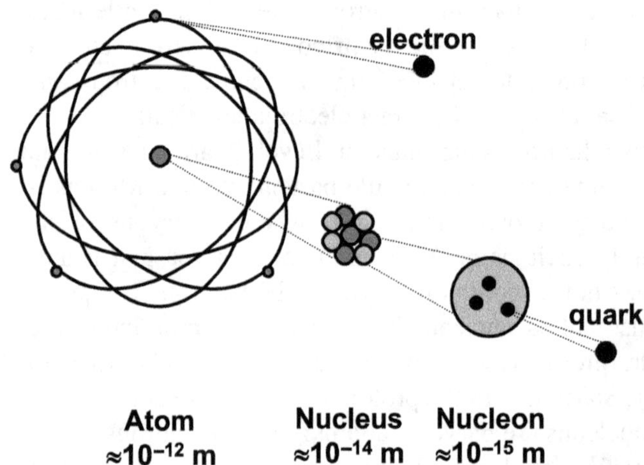

Figure 12.2. Constituents of the Atom. The Relative Sizes in Meters at the Bottom of the Figure are Approximate.

The number of protons in the nucleus determines the identity of the atom as an element. An electrically neutral atom must have as many negatively charged electrons in its electron cloud as it does positively charged protons in its nucleus. Negatively charged electrons occupy regions of space surrounding the nucleus and are bound to the positively charged nucleus by the electromagnetic interaction. The actual location of each electron is not well defined because of Heisenberg's uncertainty principle. Instead, the rules of quantum mechanics allow us to calculate the probability of finding an electron at a particular spatial location. The term 'electron cloud' denotes both our ignorance of exact spatial locations and our knowledge of probability distributions. The shape of the probability distribution is the shape of the electron cloud. Typical shapes of electron clouds are depicted as dashed lines in Figure 12.3.

Atoms and quantum numbers

Atoms are characterized by four quantum numbers. The principal quantum number characterizes the average distance of the electron from the nucleus. An increase in the principal quantum number corresponds to an increase in the average distance of the electron from the nucleus.

The secondary quantum number is represented by letters from traditional spectroscopic notation. The first few letters s, p, d, and f signify secondary quantum numbers with absolute values of 0, 1, 2, and 3, respectively. Examples of s, p, and d orbitals are sketched in Figure 12.3.

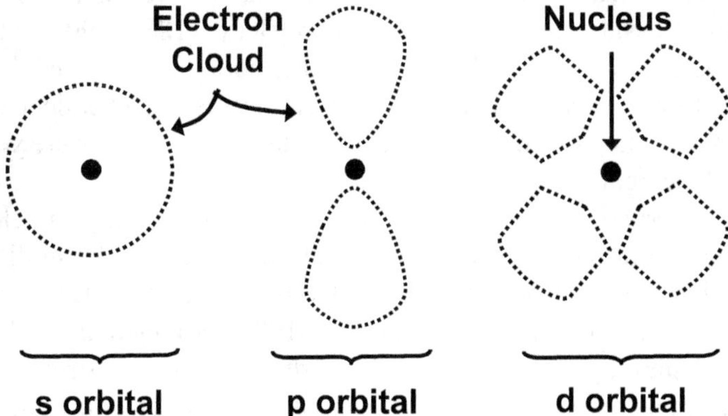

Figure 12.3. Typical Shapes of Electron Clouds.

The secondary quantum number defines the shape of the electron orbital. The orientation of the electron cloud in space is characterized by a third quantum number called the magnetic quantum number. The fourth quantum number is electron spin.

Each electron in an atom has a unique set of quantum numbers. Atomic electrons fill energy levels with the lowest energy first, and only two electrons are allowed in each energy level. The restriction to two electrons is based on the observation that electrons in magnetic fields display intrinsic electron spin.

German physicist Otto Stern (1888–1969) and countryman Walther Gerlach (1889–1979) experimentally demonstrated the existence of electron spin in 1922. They directed a beam of silver atoms through an inhomogeneous (spatially varying) magnetic field and observed the beam split into two beams. The Stern–Gerlach results were explained by Dutch-American physicists Samuel A. Goudsmit (1902–1978) and George E. Uhlenbeck (1900–1988) using electron spin. The two observed values of electron spin in the Stern–Gerlach experiment are referred to by such names as spin up and spin down; clockwise and counterclockwise spin; and helicity. Helicity defines the direction of particle spin relative to the direction of motion of the particle.

The quark model

The electron, proton, and neutron were sufficient for explaining the results of most experiments for decades. For example, the electron, proton, and neutron were used to populate the periodic table of elements first published by Russian chemist Dmitri Mendeleev (1834–1907) in 1869 (Mendeleev, 1869). Mendeleev proposed a periodic law that classified elements based on atomic weights and similarities of chemical properties.[2] His periodic table of the elements contained gaps that were yet to be identified elements.

One observation that could not be explained using only the electron, proton, and neutron was the existence of stable nuclei. An undisturbed nucleus that does not decay into smaller fragments is a stable nucleus. Japanese physicist Hideki Yukawa (1907–1981) provided an explanation of nuclear stability in 1935 when he postulated that the strong nuclear force was transmitted between nucleons by the creation and exchange of a previously undiscovered particle now called the pi meson or pion for short. A particle having many of the properties of Yukawa's pion was discovered in 1936. Further study of the properties of the observed particle, now known to be a muon, showed that it did not interact strongly with other matter. Yukawa's pion was discovered in 1947 by British physicist Cecil F. Powell (1903–1969). Both the muon and the pion were observed in cosmic ray experiments.

The existence of a burgeoning number of particles prompted people to look for particles of a more elementary nature. The quark model was introduced in 1964 by American physicist Murray Gell-Mann (1929–2019) and independently by Russian physicist George Zweig (1937–). Gell-Mann called the more elementary particle a 'quark' after James Joyce's (1882–1941) 1939 novel *Finnegan's Wake*. The word quark was used in Joyce's phrase "Three Quarks for Muster Mark" (Joyce, 1939, p. 383; Brennan, 1997, Chapter 8).

A relatively small number of quarks in the quark model could be used to construct mesons and baryons introduced in Table 12.1. Mesons are composed of a quark and an antiquark. By contrast, baryons are composed of three quarks. Today, the number of quarks and leptons has grown from the original two quarks and two leptons to six 'flavors' of quarks and six kinds of leptons shown in Figure 12.4. Each box in Figure 12.4 follows

Spin and Charge

	Three Generations of Fermion Matter		
	I	II	III
quarks	up	charm	top
	$u_{1/2}^{2/3}$	$c_{1/2}^{2/3}$	$t_{1/2}^{2/3}$
	down	strange	bottom
	$d_{1/2}^{-1/3}$	$s_{1/2}^{-1/3}$	$b_{1/2}^{-1/3}$
leptons	electron	muon	tau
	$e_{1/2}^{-1}$	$\mu_{1/2}^{-1}$	$\tau_{1/2}^{-1}$
	e-neutrino	μ-neutrino	τ-neutrino
	$\nu_{e_{1/2}}^{-1/3}$	$\nu_{\mu_{1/2}}^{-1/3}$	$\nu_{\tau_{1/2}}^{-1/3}$

gauge bosons
gluon
g_1^0
photon
γ_1^0
Z boson
Z_1^0
W boson
$W_1^{\pm 1}$

name
X_{spin}^{charge}

Higgs boson
H_0^0

Figure 12.4. Elementary Particle Properties — Charge and Spin.

the format of the box on the lower left-hand side. The symbol of the charge is denoted by *X*, the superscript charge refers to electric charge, and the subscript spin refers to particle spin. Fermions have half-integral spin, and bosons have integral spin. The fermion particles have antiparticles, which are not shown here.

Particle masses, to the extent that they are known, are shown as approximate energy values in Figure 12.5. Particle masses are expressed as energy in eV (electron volts), MeV (million eV), or GeV (Giga eV). The gluon and photon are massless. Observed masses along a row of fermions seem to increase from one generation to the next.

Table 12.2 lists constituents of matter in terms of quarks and leptons. Everyday matter that is made up of atoms is composed of up and down

	Three Generations of Fermion Matter			Masses (as Energy)	
	I	II	III		
quarks	**up**	**charm**	**top**	**gluon**	gauge bosons
	2.3 MeV	1.275 GeV	173.07 GeV	0	
	down	**strange**	**bottom**	**photon**	
	4.8 MeV	95 MeV	4.18 GeV	0	
leptons	**electron**	**muon**	**tau**	**Z boson**	
	0.511 MeV	105.7 MeV	1.777 GeV	91.2 GeV	
	e-neutrino	**μ-neutrino**	**τ-neutrino**	**W boson**	
	< 2.2 eV	< 0.17 MeV	< 15.5 MeV	80.4 GeV	
	name			**Higgs boson**	
	mass			126 GeV	

Figure 12.5. Elementary Particle Properties — Masses.

Table 12.2. Constituents of Matter.

Matter	Quarks	Leptons
Everyday	Up, down	Electron, e-neutrino
High energy	Strange, charm, bottom, top	Muon, mu-neutrino, tau, tau-neutrino

quarks, and two leptons. The up and down quarks have different electric charges and different masses. According to the quark model, a proton contains two up quarks and a down quark, while a neutron contains one up quark and two down quarks. The leptons are electrons and electron-neutrinos (e-neutrino). The remaining four quarks and leptons appear in more exotic matter. The electron, muon, and tau leptons are electrically charged, while neutrinos are electrically neutral. Strong, electromagnetic, and weak interactions are mediated by the exchange of a gauge boson (photon, gluon, or W, Z boson) between two fermions (lepton or quark).

Several attempts have been made to test or falsify the quark model. One important attempt was based on the observation that quarks are thought to have a fractional electric charge compared to the electric charge of electrons and protons (Figure 12.4). Historically, the magnitude of the electric charge of the electron has been considered the elementary charge because it matches the magnitude of the observed charge of all charged particles. The postulated fractional electric charge of a quark provides a means of searching for and identifying quarks. To date, no one has observed a particle with an electric charge that is a fraction of the magnitude of the electron charge.

The evidence for quarks does not depend solely on a search for fractional elementary charge. High-energy scattering experiments have shown that nucleons exhibit an internal structure. The internal structure can be understood if we think of nucleons as composites of quarks and treat the quarks as point particles.

Hadrons and baryons are composed of quarks. Quarks seem to bind together as particle–antiparticle pairs or triplets. Efforts to experimentally observe unbound quarks have failed so far. If quarks exist, which appears to be the case based on available experimental data, why have unbound, or free, quarks not been observed?

A mechanism called asymptotic freedom is considered the most plausible explanation for the failure to observe free quarks. Asymptotic freedom says that the interaction between quarks decreases as the separation between bound quarks decreases. Conversely, the strength of the interaction between quarks increases as quarks separate. If asymptotic freedom is correct, experiments capable of scattering a particle from the quark system should show quarks behaving like free particles within the region of space they occupy. This type of behavior is analogous to the behavior of two balls bound together by an elastic band. If the balls are near each other, the elastic band is slack and the balls behave as if they are free.

If we pull the balls apart, the elastic band resists their separation. The only way to completely separate the balls is to break the elastic band. By analogy, quarks can only be separated by disrupting the strong force binding them.

To date, experiments have shown quarks behaving like free particles inside the hadron. Experiments have also shown that high-energy electrons can scatter off quarks in a proton. If the quark acquires enough energy from the interaction, the strong interaction is broken and the quark can be ejected from the proton. The ejected energetic quark and release of energy that occurs when the strong interaction is broken can produce a narrow cone of hadrons and other particles that appears as a jet of particles.[3]

12.3 Relativistic Quantum Theory

The Standard Model of Particle Physics combined the quark model with a relativistic treatment of quantum theory. Quantum mechanics was applicable to atomic systems, and special relativity was applicable to objects moving at speeds approaching the speed of light. The two theories — quantum mechanics and special relativity — were first combined in the 1920s to form a theory now known as relativistic quantum theory.

Relativistic quantum theory had to reconcile ostensibly incompatible observations. For example, special relativity says that energy can be converted into mass and mass converted into energy in a relativistic system. On the other hand, the number of particles in a nonrelativistic quantum mechanical system is conserved, that is, the number of particles does not change. These observations implied that conservation laws in one theoretical context (quantum mechanics) were not necessarily true in another theoretical context (special relativity). Could the inconsistency be reconciled in a broader theoretical context?

A measurable physical quantity can be considered conserved if it does not change in a closed system. The validity of conservation laws can change as warranted by new observations. Conservation of energy says that the energy of a closed system does not change with time. It was considered valid until scientists observed that the number and type of particles can change when particles are created or destroyed. The correct conserved property in nonrelativistic and relativistic systems is the combination of mass and energy which can be written as mass–energy.

An important task of contemporary high-energy physics is the identification of conservation laws.

Relativistic quantum theory is designed to allow mass–energy conversion and conservation of mass–energy. The Standard Model of Particle Physics discussed below contains the quark model and is the most widely accepted relativistic quantum theory.

12.4 Standard Model of Particle Physics

Results of particle experiments and a selection of theoretical attempts to understand the results are periodically compiled and published by the Particle Data Group (PDG) (Zyla *et al.*, 2020). PDG publications include results of precision tests of a theory of the interaction of electrons and photons called quantum electrodynamics (QED). Precision tests show that QED agrees with measurements to an accuracy of one part per billion or better. QED is one of the most accurate physical theories ever developed for studying scattering systems. It is an example of a quantum field theory, that is, a theory formulated in terms of fields that have specific values at every point in spacetime, even if that value is zero.

The Standard Model of Particle Physics is a quantum field theory of strong and electroweak interactions. It contains QED as a component theory, and it assumes the existence of quarks, leptons, and the Higgs particle listed in Table 12.3. The particles have properties shown in Figures 12.4 and 12.5. Quarks and leptons are fermions with half-integral spin. Fermions cannot simultaneously occupy the same quantum state, which was originally proposed by Austrian physicist Wolfgang Pauli

Table 12.3. Fundamental Particles in the Standard Model of Particle Physics.

Fermions (Matter Particles)				**Gauge Bosons (Force Carriers)**	**Higgs Boson (Origin of Mass)**
	Generation				
	I	II	III		
Quarks	u	c	t	Photon γ	Higgs H
	d	s	b	Gluon g	
Leptons	ν_e	ν_μ	ν_τ	Z^0 boson	
	e	μ	τ	$W^\pm$ boson	

(1900–1958) and is now known as the Pauli exclusion principle.[4] Bosons have integral spin and can simultaneously occupy the same quantum state.

Fermions are considered matter particles; gauge bosons are considered force carriers; and the Higgs boson is considered essential to the origin of particle mass. The Higgs boson is named after British physicist Peter Higgs (1929–), who hypothesized that matter particles acquire mass by interacting with the Higgs boson. The Higgs boson was discovered in 2012 (CMS Collaboration, 2012). Interactions between fermions (matter particles) are mediated by gauge bosons (force carriers). Figure 12.1 illustrates some of these interactions. The photon mediates the electromagnetic interaction; the gluon mediates the strong interaction; and the weak interaction is mediated by the $W^{\pm}$ and Z^0 bosons. The as yet undiscovered graviton mediates the gravitational interaction.

Particles or fields?

A conceptual conflict arises when we consider viewing matter as either particles or fields. Thus far we have treated particles as real, physically indivisible objects. The experimental success of quantum field theories raises the question: should we be working with particles or fields? Do we need wave–particle duality, or is it sufficient to treat particles as fields?

We have two views to consider. To be specific, suppose an electron is interacting with a photon. If we adopt the particle view, then the electron is a particle, and the electromagnetic field arises from a collection of photons. On the other hand, if we adopt the field view, then each photon arises from quantization of the electromagnetic field, and each electron arises from quantization of a matter field.

The field theory view is that particles should be considered quantized excitations of fields. Each of the ‘particles’ shown in Table 12.3 should be viewed as an excitation of the corresponding field. For example, the electron is an excitation of the electron field, and the Higgs boson is an excitation of the Higgs boson field. A particle interaction occurs when there is a coupling between fields. Force carriers are excitations that mediate interactions.

Particle mass is a measure of the energy needed to make the field vibrate. A more massive field can decay into less massive fields. The excitation of a field is shown in Figure 12.6(A), and the decay of a more massive field into two less massive fields is depicted in Figure 12.6(B).

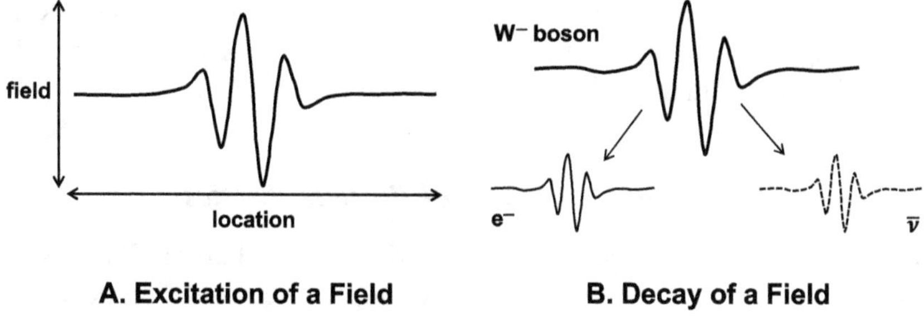

Figure 12.6. Excitation and Decay of a Field.

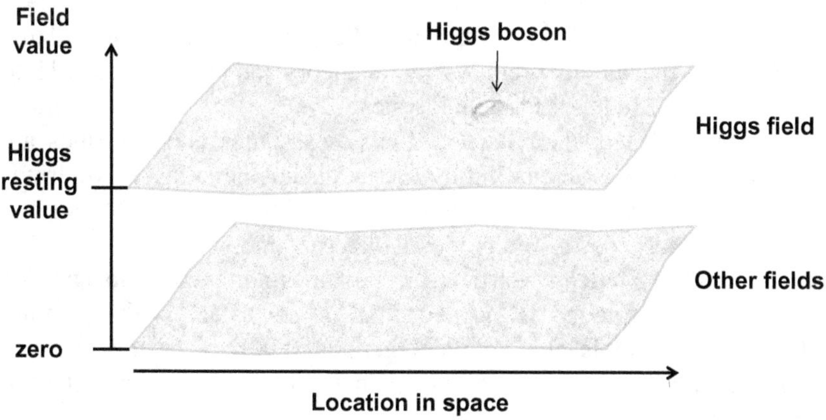

Figure 12.7. The Higgs Field (Modified from Carroll, 2012, p. 34).

Why do we appear to observe distinct particles with mass if matter is the excitation of a field? The Higgs field provides a field theory explanation of the origin of mass. The resting value of the Higgs field in empty space is different from all other fields because it is nonzero. The resting value illustrated in Figure 12.7 is also known as the vacuum expectation value. The resting value of other fields is zero. All of the fields except the Higgs field would be massless and move at the speed of light if the Higgs field was not present. We noted above that the Higgs boson was discovered in 2012 (CMS Collaboration, 2012).

In the presence of the Higgs field, other fields, except for the photon and possibly the neutrino, acquire mass by coupling to the Higgs field.

The Standard Model of Particle Physics treats the neutrino as massless even though massive neutrinos have been observed. A Higgs boson is depicted as an excitation of the Higgs field in Figure 12.7.

Anomalies of the Standard Model of Particle Physics

Despite its record of success, the Standard Model of Particle Physics is associated with many anomalies. A few examples are listed as follows:

- The mathematical equations of quantum field theory are difficult to solve when applied to bound-state systems such as the behavior of electrons bound to an atomic nucleus.
- Masses of quarks, leptons, and the Higgs particle are some of the parameters that are not predicted by the theory and must be entered into the Standard Model of Particle Physics.
- The Standard Model of Particle Physics assumes that neutrinos are massless, yet experiments imply that neutrinos have mass. Neutrinos with mass can be included in an extension of the Standard Model, but this requires the use of more unexplained parameters.
- The Standard Model of Particle Physics does not account for gravity. A theory of gravity is needed that is consistent with quantum field theory and is able to yield general relativity as the classical approximation. The graviton shown in Figure 12.1 has not yet been observed.
- The Standard Model of Particle Physics does not explicitly account for dark matter and dark energy.

Proponents of the Standard Model of Particle Physics recognize that anomalies may change the mainstream paradigm.

# # #

The Standard Model of Particle Physics, general relativity, and cosmology range in scale from the subatomic to the universal. General relativity and cosmology provide many of the basic tools needed to explain the expansion of the universe. The Standard Model of Particle Physics provides mechanisms for explaining the initiation of the universe in terms of particle interactions and quantum effects. For example, the vacuum process shown in Figure 11.4(C) may have played a role in the initiation of the Big Bang. We describe how these ideas are incorporated into the leading theory of universal creation and evolution in the next chapter.

Endnotes

1. Sources on the Standard Model of Particle Physics include Veltman (2018), Ecker (2019), Schmitz (2019), Zyla *et al.* (2020), and Weinberg (2021).
2. Sources about the periodic table include Mendeleev (1869, 1889, 1905), Teerikorpi *et al.* (2019), and Edwards *et al.* (2020).
3. For a more detailed conceptual discussion of jets and asymptotic freedom, see Schmitz (2019).
4. Sources on the Pauli exclusion principle include Weinberg (2015), Veltman (2018), and Ecker (2019).

Chapter 13

The Inflationary Universe

Many cosmological models have been proposed as universal scale models of the aftermath of the Big Bang.[1] Albert Einstein (1879–1955) solved the equations of the general theory of relativity for a static, unchanging universe. Russian mathematician Alexander Friedmann (1888–1925) solved Einstein's equations in 1922 for a model of a dynamic universe, that is, a universe that could change with time. Belgian cosmologist and priest Georges Lemaître (1894–1966) independently developed a dynamic universe model from Einstein's equations in 1927. Lemaître's model universe could be static, expand, or contract. By 1931, Edwin P. Hubble (1889–1953) had published his velocity–distance relation and suggested the idea of an expanding universe.

Lemaître was the first to propose the idea that the universe began in a moment of creation. He proposed the hypothesis of the primeval atom in 1931 and used the primeval atom to initiate an explosive expansion of the universe.

By the 1940s, Russian–American astrophysicist George Gamow (1904–1968) applied Friedmann's and Lemaître's dynamic solutions of Einstein's equations to study the behavior of matter in the early universe. He predicted the existence of a cosmic background radiation that was not observed until 1964 by Penzias and Wilson.

British astronomer Fred Hoyle (1915–2001) first used the term Big Bang on a 1950 BBC radio broadcast to describe the apparently explosive expansion of the universe. Hoyle appeared to use the term derisively as he argued for a steady-state universe (Singh, 2004, pp. 352–353). The steady-state theory was based on the hypothesis that matter was being continuously created in the expanding universe to preserve the homogeneity and isotropy of matter distributed throughout the universe. In this theory, the universe had no beginning and no end.

Fred Hoyle (1948), Hermann Bondi, and Thomas Gold (Bondi and Gold, 1948) proposed the modern steady-state theory. Measurements of the distribution of matter have shown that the density of matter is decreasing as the universe expands. This observation and the observation of the cosmic microwave background (CMB) radiation do not support the steady-state universe hypothesis. Danish historian Helge Kragh (1944–) presented an analysis of the historical competition between the Friedmann–Lemaître–Gamow Big Bang universe and the Hoyle–Bondi–Gold steady-state universe (Kragh, 1996).

The purpose of this chapter is to use what we have learned to describe the inflationary model of cosmology, which is the leading scientific explanation of the Big Bang and its aftermath. The inflationary universe is initiated by a quantum process on the subatomic scale and expands in curved spacetime. We begin with the moment of creation.

13.1 The Big Bang — The Moment of Creation

The cosmological interpretation of Hubble's law seemed to favor the idea of a universe that was expanding from a much more compact accumulation of mass and energy. The primeval atom hypothesized by Georges Lemaître in 1931 provided a mechanism for initiating an explosive expansion of the universe that began in a moment of creation. Other unproven hypotheses have been proposed to explain what happened at the moment of the Big Bang. One hypothesis is that all of the mass–energy of the universe was initially a hot, dense state located at a point in spacetime called a singularity. This idea is similar to Lemaître's primeval atom. The source of the initial mass–energy is not known. Here we focus on the possibility that the entire universe is the outgrowth of a quantum process.

Quantum fluctuations are a natural mechanism for explaining the origin of the universe.[2] A quantum fluctuation is a short duration, random change in energy at a point in space. The duration and change in energy are constrained by the energy–time uncertainty relation. Examples of quantum fluctuations[3] include the vacuum bubble and the vacuum polarization shown in Figure 13.1.

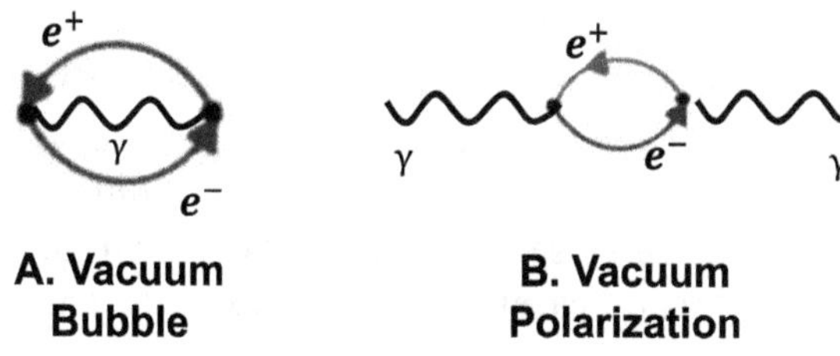

Figure 13.1. Examples of Particle Interactions.

The vacuum bubble in Figure 13.1(A) shows an electron e^-, a positron e^+, and a photon γ materialize from the vacuum at the left-hand vertex and annihilate one another at the right-hand vertex. The vacuum bubble exists for a period of time consistent with the uncertainty relation for energy and time. The vacuum bubble is characterized by no external lines.

Vacuum polarization in Figure 13.1(B) shows a photon creating an electron–positron pair at the left-hand vertex. The photon is recreated at the right-hand vertex when the electron and positron annihilate each other. The electron and positron exist for a period of time consistent with the uncertainty relation for energy and time. Vacuum polarization has been shown to contribute to a slight shift in atomic energy levels.

Quantum fluctuations in vacuum can distort spacetime and produce a locally dense accumulation of particles known as a quantum foam. The quantum foam consists of several quantum fluctuations in a region of space and duration of time constrained by uncertainty relations. If a fluctuation can sustain itself, it becomes a source of matter that can impart a local expansion of space. An expansion of space creates more room for additional fluctuations in a snow-balling expansion of the universe. According to this view, the Big Bang can be initiated by a quantum foam.

13.2 The Standard Model of Cosmology

Open universe and closed universe models are illustrated in Figure 11.1. Open universe models assume the universe began as an explosive release of mass–energy and the indefinite expansion of spacetime. Time and space began with the Big Bang, and it can be argued that spacetime did not exist prior to the Big Bang. The geometry of spacetime depended on the existence of mass–energy, which possibly began as a quantum foam.

Another view is provided by closed universe models. Spacetime expansion in closed universe models eventually stops and reverses itself until all of the mass–energy of the universe is compressed into a new 'primeval atom'. If another Big Bang occurs, spacetime again starts to expand. In oscillating universe models, the expansion and contraction phases periodically repeat. Spacetime has existed as long as mass–energy has existed.

Historically the leading cosmological models of the mid-20th century were the Big Bang model and the steady-state model. Today, the two most influential cosmological models are the standard model of cosmology and an inflationary model called the Lambda-cold dark matter (ΛCDM) model. Figure 13.2 presents a comparison of the expansion of the standard model and the expansion of the inflationary model. The inflationary model begins with a smaller radius than the standard model and rapidly expands during an inflation era until it matches the expansion of the standard model.

Inflationary expansion of spacetime exceeds the speed of light. Einstein's theory of special relativity showed that matter cannot travel faster than the speed of light. If true, how did spacetime expand faster than the speed of light? We pointed out in Chapter 10 that the speed of light limits the speed of an object within spacetime; it does not limit the rate of expansion of spacetime itself.

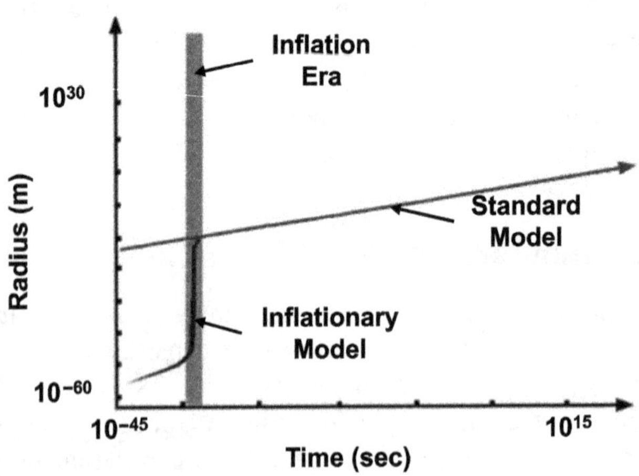

Figure 13.2. Inflationary Expansion of the Universe (Modified from Guth, 2002).

Table 13.1. The Standard Model of Cosmology.

Event	Comment
Big Bang	Origin of CMB
Hubble expansion	Friedmann–Lemaître expansion of spacetime Ordinary matter is only type of matter

According to the standard model of cosmology summarized in Table 13.1, the universe began with a Big Bang, and spacetime is continuing a Friedmann–Lemaître expansion identified as the Standard Model in Figure 13.2. The standard model of cosmology assumes that ordinary matter was the only known type of matter, and the universe obeys the cosmological principle, that is, the spatial distribution of matter is homogeneous and isotropic.

The standard model of cosmology was able to account for universal expansion, but other issues remained and pointed to the need for a revised cosmological model. Some of the issues that motivated revisions to the standard model of cosmology are considered next.

13.3 Cosmological Model Parameters

The universe appears to be expanding as the aftermath of a 'Big Bang', or primordial explosion. Big Bang models are based on measurements of astronomical distances, observations of the cosmological redshift, the cosmic background radiation, and the distribution of matter in the universe.[4]

The general theory of relativity provides the mathematics of modern cosmological models. The mathematics are simplified by adopting the Cosmological Principle, which assumes matter in the universe is homogeneous and isotropic. The assumption of homogeneity says that matter in the universe is uniformly distributed on some large scale, and the assumption of isotropy says that the distribution of matter does not depend on direction. The assumptions of homogeneity and isotropy are subject to experimental testing.

The isotropy of matter in the universe is determined by measuring the cosmic background radiation. The universe is anisotropic if the distribution of matter depends on the direction we look. The isotropy of the universe is an important measurable parameter of cosmological models.

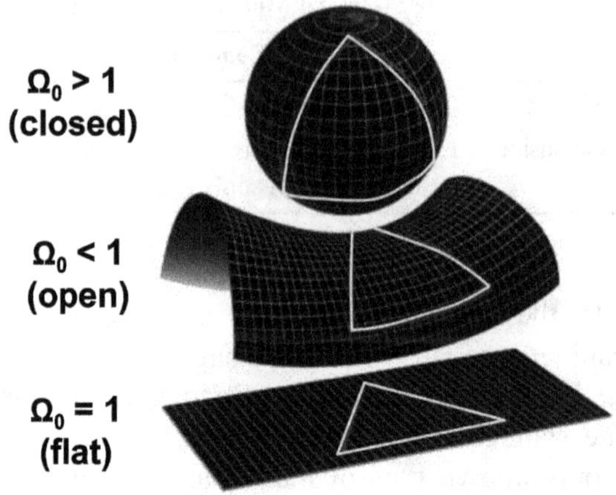

Figure 13.3. Geometry of Possible Universes (NASA Geometry, 2020).

The standard model of cosmology depends on three observable parameters: the Hubble expansion rate, a dimensionless deceleration parameter, and the critical density of the universe. The Hubble expansion rate was discussed previously. The deceleration parameter is estimated from the relationship between the magnitude of the redshifts of E-type galaxies and their distances from Earth. E-type galaxies tend to be elliptical galaxies and can serve as standard candles.

The critical density of the universe is the third cosmological parameter. Critical density is the density of the flat universe shown in Figure 13.3. The symbol Ω_0 (called omega zero) is the ratio of observed density to critical density. The universe is flat if the observed density equals the critical density, $\Omega_0 = 1$. The universe is open if Ω_0 is less than 1, $\Omega_0 < 1$. The universe is closed if Ω_0 is greater than 1, $\Omega_0 > 1$.

A density measurement tells us whether the cosmological model is open, closed, or flat. The observed density is the observable matter of the universe divided by the estimated volume of the universe. Several factors can affect the value of observed density.

Estimating the size of the universe

The radius of the universe can be estimated by dividing the rate of expansion of the universe by the maximum speed of recession of a receding galaxy. The rate of expansion of the universe is based on measurements of the cosmological redshift. If we assume the rate of expansion is given by Hubble's law (also known as the Hubble–Lemaître law), the galaxies are receding from Earth at speeds proportional to their distance. The size of the universe depends on a given geometry.

Hubble flow and cosmic time

A cosmic time can be defined in terms of Hubble's law. The motion of galaxies that satisfy Hubble's law is called Hubble flow and corresponds to motion in a homogeneous, expanding universe. Cosmic time is measured using clocks moving with the Hubble flow. The timeline of events since the Big Bang is expressed here in terms of cosmic time.

Estimating the amount of matter in the universe

The amount of matter in the universe is estimated by summing observable matter such as galaxies and nebulae. Estimates of universal matter ordinarily assume the validity of the Cosmological Principle, which says that the distribution of matter in the universe is uniform (homogeneous) and independent of the direction in which we look (isotropic). If we look at the distribution of matter near us, such as our Solar System, the Milky Way galaxy, or the Local Group of galaxies, we do not see a uniform distribution of matter. The Cosmological Principle does not apply on this scale because the gravitational influence of astronomical objects on one another in the Local Group disrupts the hypothesized homogeneity and isotropy. We must look at an even larger region of space before we begin to observe a homogeneous and isotropic distribution of matter.

The Cosmological Principle seems applicable to the region of the universe we can observe. Does the Cosmological Principle apply everywhere? Research organizations like NASA are studying this question (NASA-WMAP Science Team, 2022).

One source of error in our estimate of the size of the universe is the assumption that we have accurately measured all of the mass in the universe. A relatively recent challenge to this assumption is the observation that the neutrino has mass. The neutrino in the Standard Model of Particle Physics is supposed to be massless. It was originally introduced into physics to preserve the laws of conservation of energy and momentum.

Experiments have shown that the neutrino has a small but nonzero rest mass. The abundance of massive neutrinos in the universe could contribute to the density of the universe, but research to date suggests that the mass of the neutrino is too small to account for all of the unobserved matter that is inferred from studies of the motion and structure of galaxies.

Black holes are another possible source of underestimated mass. According to theory, the gravitational field of a black hole is so strong that light cannot escape from its gravitational pull by classical methods. Particles can escape from black holes by quantum tunneling, a quantum mechanical phenomenon. Light that escapes from a black hole provides a method for detecting black holes. Several astronomical objects have many of the properties people expect a black hole to have.[5] We do not know how many black holes there are. Consequently, their mass, which could be substantial, is probably underestimated.

13.4 Dark Matter

In addition to the above sources of error in our estimate of mass in the universe, we must correct for unobservable dark matter. Ordinary matter is the luminous matter we see in the universe. Observations of the distribution and motion of luminous mass in the universe implied the need for a new form of matter.[6] California Institute of Technology astrophysicist Fritz Zwicky (1898–1974) found that the motion of galaxies in the Coma cluster of galaxies was not adequately explained by the observable matter (Zwicky, 1933, 1937). It appeared that there must be unobserved matter that Zwicky called ‘dark matter’.

Zwicky’s observations were confirmed by American astrophysicist Vera Rubin (1928–2016) and colleagues. They studied the motion of spiral galaxies (Bennet *et al.*, 2018) in the 1960s and 1970s. Stars at the fringes of the Andromeda Galaxy moved faster than expected based on the gravitational attraction associated with the mass of luminous stars in the galaxy. It appeared that more matter was needed to explain the motion of stars at the fringes of the Andromeda Galaxy.

Working with colleague Kent Ford, Rubin (Rubin *et al.*, 1976a, 1976b) measured “orbital speeds of hydrogen gas clouds in many other spiral galaxies (by studying Doppler shifts in the spectra of hydrogen gas) and discovered that the behavior seen in Andromeda is common. Although Rubin and Ford did not immediately recognize the significance of the results, they were soon arguing that the universe must contain substantial quantities of dark matter" (Bennet *et al.*, 2018, p. 472). By the 1980s, the evidence for dark matter implied that our best theory of gravity was wrong, or else dark matter must exist.

Dark matter and ordinary matter interact gravitationally and seem to be present in approximately the same regions of space. Stated differently, dark matter does not appear to be present in regions of space without visible galaxies. We can detect ordinary matter when it interacts with electromagnetic radiation (light). Dark matter seems to interact only weakly or not at all with electromagnetic radiation. Some physicists have postulated that dark matter is a new type of particle called a WIMP (weakly interacting massive particle) or astronomical object called a MACHO (massive compact halo object).

The existence of dark matter is implied by observing its apparent gravitational influence on the distribution of ordinary matter. These observations can be explained if matter consists of approximately 15–20% ordinary matter and 80–85% dark matter. Dark matter remains a mystery to science, and its existence implies that most of the matter in the universe is a mystery. A related mystery is dark energy.

13.5 Dark Energy and Inflation

Figure 13.4 is a sketch of Hubble's law. The solid line in the figure illustrates the linear relationship associated with relatively small redshift galaxies. The three broken lines (dots and dashes) at higher redshift values represent the part of the relationship that needed to be measured at the end of the 20th century. In 1998, two independent teams of astronomers tried to extend Edwin Hubble's law of universal expansion by measuring the distance to supernovae in galaxies with larger redshifts than observed previously. The two teams thought they would observe a decelerating

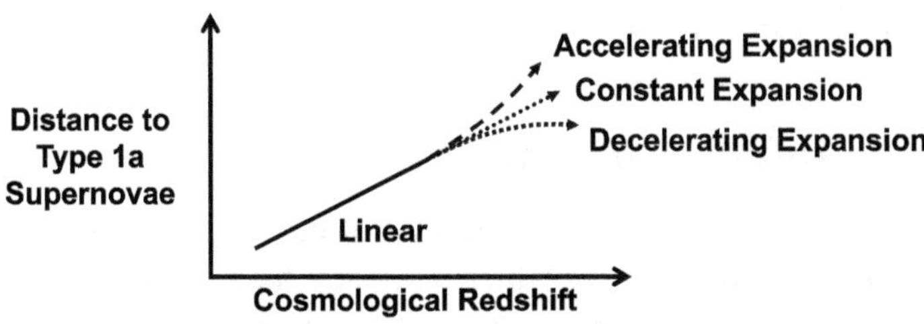

Figure 13.4. Extension of Hubble's Law.

expansion. Instead, they found that the expansion of the universe followed the accelerating expansion represented by the upper dashed curve. Observations showed that the universe is expanding at an increasing rate. What is causing this accelerating expansion?

The accelerating expansion is attributed to a form of energy called dark energy. Dark energy is included in cosmological models by adding a term called the cosmological constant to a general relativistic equation that relates matter and the geometry of spacetime. When viewed within the context of general relativity, dark energy appears to behave as a gravitationally repulsive form of energy that exerts a negative pressure.

Einstein first introduced the cosmological constant in 1917 in an effort to explain a universe which was then considered unchanging, only to learn later from Hubble's measurements that the universe was expanding. Einstein considered the cosmological constant a major blunder.[6] The cosmological constant has been resurrected to help explain observations that the expansion rate of the universe is accelerating. The cosmological constant is often represented by the Greek letter Λ (lambda).

Dark energy is represented by the cosmological constant if dark energy is constant in time and space. On the other hand, dark energy is represented by a variable called quintessence if dark energy is a function of time and space. The term quintessence means fifth essence. It refers to the fifth element of the ancient Greeks, who thought that the elements were earth, air, fire, water, and ether. Ether filled the universe beyond Earth. The dependence of quintessence on space and time appears in an equation of state that relates pressure and density.

Dark energy accounts for approximately 70% of the cosmic energy density. CDM accounts for approximately 25% of the energy density of the universe. The mass and energy that we observe as ordinary matter is the remaining 5% of the energy density of the universe. If we combine dark energy and dark matter, we find that approximately 95% of the mass–energy in the universe is poorly understood. The percentages are subject to revision as research progresses.

13.6 The Inflationary Universe

The procedure for calculating the ratio Ω_0 of observed density to critical density outlined above tells us whether we are in an open, closed, or flat universe. The observed density of the universe is obtained by dividing the corrected amount of matter in the universe by the size of the universe.

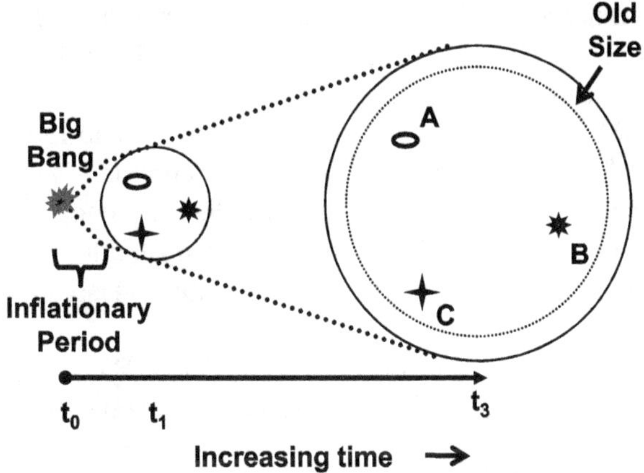

Figure 13.5. Inflation (Not to Scale).

There are several assumptions and uncertainties in the procedure. For example, what if we do not know the correct rate of expansion of the universe? Has the rate of expansion changed?

Inflationary expansion shown in Figure 13.2 suggests that the expansion rate of the universe has changed with time.[7] Figure 13.5 illustrates a period of rapid expansion immediately after the Big Bang. The expansion rate during the inflationary period exceeded the Hubble expansion rate. The 'Old Size' in the figure refers to the size of the universe without inflation. Inflationary expansion yields a bigger universe than the expansion without inflation.

Modern cosmological models incorporate observations of the cosmological redshift, the CMB, and the distribution of matter in the universe. The Wilkinson Microwave Anisotropy Probe (WMAP) has provided measurements of anisotropy in the cosmic background radiation. By comparing WMAP measurements with cosmological models, cosmologists are finding evidence to support the view that the universe experienced inflationary expansion immediately after the Big Bang. Furthermore, it appears that the expansion of the universe is accelerating. This view depends on the hypothesized existence of dark energy, which is a poorly understood type of energy.

13.7 The ΛCDM Model

The ΛCDM cosmological model is an inflationary cosmological model. It includes an inflationary period of rapid spacetime expansion shortly after the Big Bang and before the Friedmann–Lemaître expansion period associated with the Hubble expansion rate. Λ refers to the cosmological constant introduced by Albert Einstein.

The ΛCDM model is summarized in Table 13.2 and Figure 13.2. The dark matter halo referred to in the table is an accumulation of dark matter. Ordinary matter is present in the dark matter halo as a hot, gaseous phase. Within the context of the ΛCDM model, ordinary matter stars and galaxies form inside the dark matter halo as the gaseous phase of ordinary matter cools.

Figure 13.6 illustrates several features of Big Bang expansion. The present is depicted by the WMAP spacecraft shown on the right-hand side

Table 13.2. The ΛCDM Cosmological Model.

Event	Comment
Big Bang	Origin of cosmic background radiation
Inflation	Rapid expansion of spacetime caused by dark energy (nonzero cosmological constant)
Hubble expansion	Uniform expansion of spacetime Formation of dark matter halos Formation of ordinary matter objects such as nebulae, galaxies, and star clusters

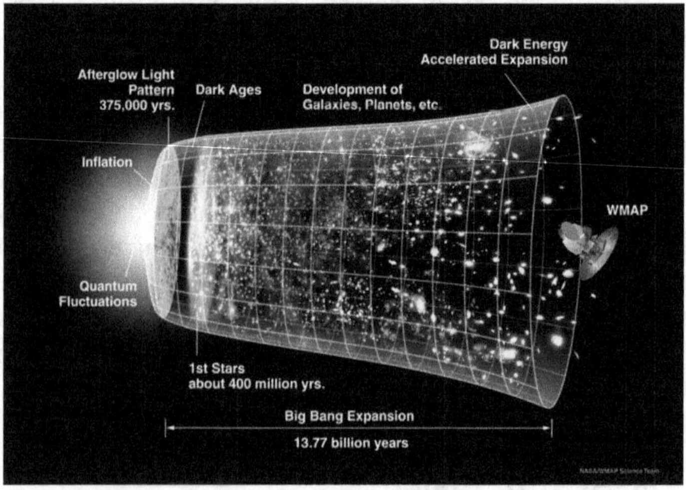

Figure 13.6. Timeline of the Universe (NASA-WMAP, 2022).

of the figure. The Big Bang is represented by quantum fluctuations on the left-hand side of the figure. An inflationary period immediately follows the quantum fluctuations and precedes the afterglow light pattern shown approximately 375,000 years after the Big Bang. The Dark Ages following the afterglow light pattern is the period when the gravity of matter amplifies primordial fluctuations into small amplitude ripples in spacetime. Ordinary matter becomes enclosed in astronomical dark matter structures. The first stars appear about 400 million years after the Big Bang. Dark energy leads to a relatively recent accelerated expansion of the universe. According to Figure 13.6, the age of the universe is approximately 13.77 billion years after the Big Bang.

The cosmological constant problem

Einstein's original cosmological model did not include the cosmological constant. It showed the curvature of space varying in time and location. Einstein introduced the cosmological constant to serve as a repulsive force for the purpose "of making possible a quasi-static distribution of matter, as required by the fact of the small velocities of the stars" (Einstein, 1917, p. 188). The resulting cosmological model with the cosmological constant represented a static, very slowly changing, or quasi-static universe.

The cosmological constant Λ was reintroduced into the equations of general relativity to account for inflationary expansion. It has been combined with the energy density term and is described as empty space "full of a dark energy with a density and pressure that depends on the value of lambda" (Muller, 2016, p. 158). American physicist Alan Guth (1947–) was the first to postulate an energy field called an inflaton field that could provide a repulsive gravitational push

> **Q13.1**. "so forceful that the speck of space [occupied by the inflaton field] would inflate explosively, almost instantaneously stretching to as large as the observable universe." (Greene, 2020, p. 52)

A theoretical advantage of associating Λ with dark energy is that quantum fluctuations can provide a negative pressure to explain universal inflation. Energy from quantum fluctuations in vacuum is called vacuum energy.

If the hypothesis that vacuum energy drives inflation is tested,

> **Q13.2**. "we know that the dark energy that accelerates the expansion of the universe has a mass density of about 10^{-29} gram/cc, the value predicted from the quantum physics theory is 10^{+91}. The theory is wrong by a factor of 10^{120}. This disagreement has been called 'the worst theoretical prediction in the history of physics'." (Muller, 2016, p. 158)

The disagreement is thought to arise from QFT calculations that lead to unphysical infinities. The infinities result from the inclusion of quantum field vibrations in the calculations that range from very long wavelengths down to zero. Infinities can be removed if we impose a wavelength cutoff that eliminates wavelengths greater than the wavelength cutoff. An example of a wavelength cutoff is the Planck scale ℓ_{Planck}, which is approximately 10^{-35}m.

> **Q13.3**. "If we estimate the quantum contribution to the vacuum energy by imposing a Planck-scale cutoff on the allowed modes, we get a finite answer rather than an infinite one, but that answer is 10^{122} times larger than the value we actually observe." (Carroll, 2019, p. 258)

The disagreement between theory and observation is known as the cosmological constant problem or vacuum catastrophe. Inflation continues to be a topic of research.[7]

# # #

The ΛCDM model provides a plausible narrative about the origin and evolution of the physical universe. The universe seems to have originated from one of two basic options: the mass–energy of the universe was created from nothing as in an open universe model, or the mass–energy of the universe is eternal and passed through periods of expansion and contraction as in an oscillating universe model. Creation from nothing suggests that spacetime did not exist until the Big Bang. On the other hand, the existence of eternal mass–energy and an oscillating universe model suggests that spacetime is eternal.

Expanding open, closed, and flat universe models include a period of expansion that implies the universe is evolving from past to future along an as yet undefined arrow of time. The arrow of time is discussed in the next chapter.

Endnotes

1. Sources on cosmology include Einstein (1917), Pais (1982), Peebles (1993), Weinberg (1993, 2008), Kragh (1996), Liddle (1999), Singh (2004), Perlov and Vilenkin (2017), Bennett *et al.* (2018), Fraknoi *et al.* (2018), Kisslinger (2017), Odenwald (2019), Teerikorpi *et al.* (2019), Malkan and Zuckerman (2020), and Hartle (2021).
2. Several authors have speculated on the role of quantum fluctuations of the vacuum, including speculation on the creation of the cosmos from vacuum, quantum foams, and multiverses. For example, see Misner *et al.* (1973), Lindley (1987), Linde (1987), Liddle (1999), Greene (2000, 2020), Tyson and Goldsmith (2004), Davies (2007), Hawking and Mlodinow (2010), Baggott (2013), Rovelli (2017), Bennett *et al.* (2018), Carroll (2019), and Hartle (2021).
3. Sources on vacuum bubbles and vacuum polarization include Gross (1993, Sections 11.4 and 11.5), Veltman (2018, Section 9.2), and Schmitz (2019, Section 10.7).
4. Sources that discuss cosmological parameters include Misner *et al.* (1973), Peebles (1993), Weinberg (1993), Hogan (1998), Bergstrom and Goobar (1999), Liddle (1999), Ludvigsen (1999), Fukugita and Hogan (2000), Goldsmith (2000), Kolb and Turner (2000), Singh (2004), Tyson and Goldsmith (2004), Padmanabhan (2006), Bennett *et al.* (2018), Odenwald (2019), and Hartle (2021). A different approach to cosmology that does not rely on a Big Bang was advocated by Burbridge *et al.* (1999).
5. Stars and black holes are discussed in many sources, such as Misner *et al.* (1973), Peebles (1993), Blandford and Gehrels (1999), Padmanabhan (2006), Maoz (2007), Kisslinger (2017), Bennett *et al.* (2018), Fraknoi *et al.* (2018), Malkan and Zuckerman (2020), and Hartle (2021).
6. Sources that discuss dark matter and dark energy include Srednicki (2000), Bernstein *et al.* (2000), Perlmutter (2003), Tyson and Goldsmith (2004), Silk (2006), Gribbin (2006), Freeman and McNamara (2006), Schwarzschild (2007a, 2007b), Conselice (2007), Bennett *et al.* (2018), Fraknoi *et al.* (2018), and Odenwald (2019). Rubin (2006) gives a personal account of one of the earliest observations that provided evidence for dark matter.
7. Sources that discuss inflation include Linde (1987), Peebles (1993), Bergstrom and Goobar (1999), Liddle (1999), Bucher and Spergel (1999), Kolb and Turner (2000), Singh (2004), Tyson and Goldsmith (2004), Padmanabhan (2006), Conselice (2007), Muller (2016), Kisslinger (2017), Bennett *et al.* (2018), Fraknoi *et al.* (2018), Odenwald (2019), Teerikorpi *et al.* (2019), Malkan and Zuckerman (2020), and Hartle (2021).

Part 4

The Future of Time

Chapter 14

Time's Arrow

The evolution of the universe described in Chapter 13 was presented in terms of cosmic time, which is measured using clocks moving with the Hubble flow. Cosmic time provides a monotonically increasing timeline that captures our sense that the evolution of the universe is asymmetric in time: the universe is evolving from past to future and never seems to reverse.

British astrophysicist Arthur Eddington (1882–1944) recognized the ongoing flow of time in 1929 and suggested that physicists were not adequately including temporal flow in their theories:

> **Q14.1**. "The great thing about time is that it goes on. But this is an aspect of it which the physicist sometimes seems inclined to neglect." (Eddington, 1929, p. 68)

He related temporal asymmetry to randomness:

> **Q14.2**. "It is possible to find a direction of time on the four-dimensional map by a study of organization. Let us draw an arrow arbitrarily. If as we follow the arrow we find more and more of the random element in the state of the world, then the arrow is pointing towards the future; if the random element decreases the arrow points towards the past. That is the only distinction known to physics." (Eddington, 1929, p. 69)

He associated the random element with the thermodynamic concept of entropy:

Q14.3. “The practical measure of the random element which can increase in the universe but can never decrease is called entropy.” (Eddington, 1929, p. 74)

Eddington chose the term ‘time’s arrow’ to describe temporal asymmetry:

Q14.4. “I shall use the phrase ‘time’s arrow’ to express this one-way property of time which has no analogue in space.” (Eddington, 1929, p. 69)

He believed that the

Q14.5. “law that entropy always increases — the second law of thermodynamics — holds, I think, the supreme position among the laws of Nature.” (Eddington, 1929, p. 74)

Eddington’s view of time’s arrow has not been adopted by all modern physicists. American physicist Richard A. Muller (1944–) argued that Eddington’s relationship between time’s arrow and entropy was not a causal relationship. According to Muller,

Q14.6. “the only justification for it was that both entropy and time were increasing. That’s a correlation, not causation.” (Muller, 2016, p. 172)

The purpose of this chapter is to discuss time’s arrow, which is also known as the arrow of time. We begin by reviewing entropy and the second law of thermodynamics.

14.1 Entropy and Thermodynamics

Thermodynamics is the study of the flow of heat energy. The concept of heat energy evolved from the study of burning and heating.[1] English clergyman and chemist Joseph Priestley (1733–1804) thought that a substance called ‘phlogiston’ was released as a flame during the combustion process (Wolff, 1967, Chapter 2). French chemist Antoine Lavoisier (1743–1794, by guillotine) identified phlogiston as the element oxygen (Wolff, 1967, Chapter 3). Today, we know that combustion consumes oxygen when a combustible material is burned in air. Lavoisier explained the behavior of a material subjected to heating and cooling in terms of a substance called caloric. Caloric was an invisible and 'imponderable', or weightless, substance. According to Lavoisier, heated objects expand when caloric fills the space between the particles of an object. Conversely, an object contracts when it is cooled because it loses caloric. The concept of heat eventually replaced the concept of caloric.

American Benjamin Thompson (1753–1814), who became Count Rumford of Bavaria, provided experimental evidence that heat was not a conserved substance. Several people independently realized that heat was a form of energy and that energy was conserved by the 19th century. Early advocates of energy conservation included Germans Julius Mayer (1814–1878) and Hermann Ludwig von Helmholtz (1821–1894), Englishman James Joule (1818–1889), and Ludvig A. Colding (1815–1888) in Denmark. Peter Wolff[2] published an English version of Helmholtz's seminal paper on energy conservation (1965, Chapter 8).

The kinetic theory of atoms emerged when Englishman John Dalton's (1776–1844) concept of atoms (Wolff, 1967, Chapter 4; Gjertsen, 1984, Chapter 11) was combined with the concepts of heat and conservation of energy. The kinetic theory relates the temperature of an object to the motion of atoms within the object. According to the kinetic theory, a heated object expands because heating increases the kinetic energy, or energy of motion, of its atoms. Similarly, a cooled object contracts because cooling decreases the kinetic energy of atoms in the object.

The ideas of phlogiston and caloric were at the core of paradigms, or widely held beliefs, in the 17th and 18th centuries. These paradigms were eventually replaced by the kinetic theory and statistical mechanics in the 19th century.[3] Our discussion of entropy and time's arrow requires a familiarity with the laws of thermodynamics. We begin by defining fundamental concepts and then summarize the laws of thermodynamics.

Thermodynamic systems and states

The universe can be described as the combination of a system and its surroundings (Figure 14.1). The system is the part of the universe that we are considering, and everything outside of the system is the surroundings. The boundary distinguishes the system from the surroundings. Table 14.1

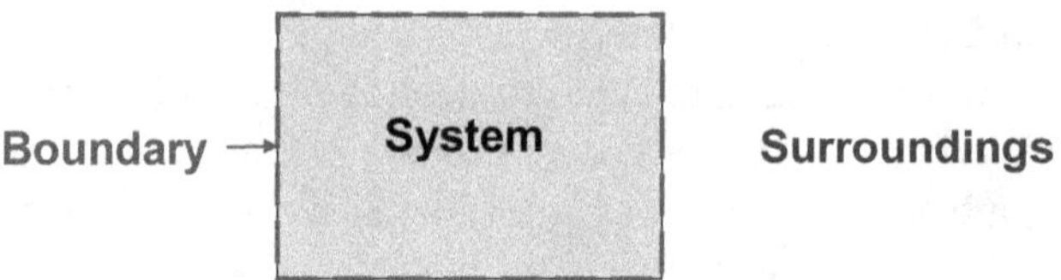

Figure 14.1. System and Surroundings.

Table 14.1. The Interaction Between System and Surroundings.

System	Interaction with Surroundings
Isolated	Does not interact
Closed	Can exchange energy but not matter
Open	Can exchange energy and matter

summarizes the interaction of the surroundings with three different types of systems.

Systems can be described from microscopic and macroscopic points of view. The microscopic approach characterizes the behavior of a large number of particles in a system using statistical variables. For example, the kinetic theory relates the temperature of an ideal gas to average values of velocity and kinetic energy of the particles comprising the ideal gas.

The macroscopic approach characterizes the behavior of the system in terms of large-scale effects of the particles in the system. Large-scale effects can be measured by instruments such as thermometers and pressure gauges. Macroscopic variables include pressure P, volume V, and temperature T.

The state of a system is a set of values of state variables that characterize the system. Pressure P, volume V, and temperature T are examples of state variables for an ideal gas. A process is a change from one state of the system to another. The system undergoes a process when any of the state variables change.

A process is illustrated in Figure 14.2. The system is a closed system containing gas. State 1 is the initial state with state variables $\{P_1, V_1, T_1, \ldots\}$. State 2 is the final state with state variables $\{P_2, V_2, T_2, \ldots\}$. The difference between the states is the result of increasing the volume of the closed system by the process of changing the location of the piston. The volume occupied by the gas increases from State 1 to State 2 with corresponding changes in pressure and temperature.

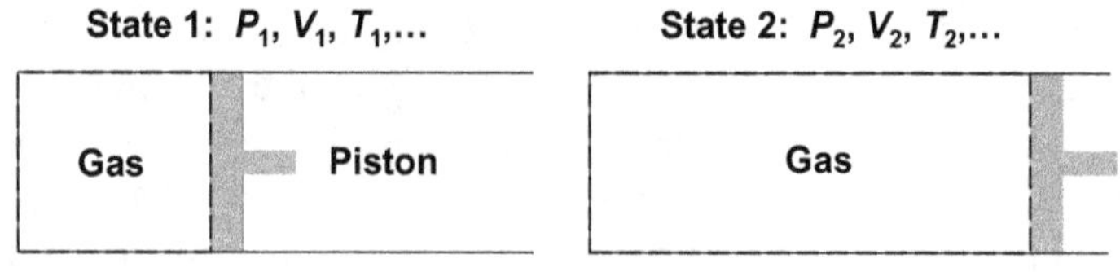

Figure 14.2. Process.

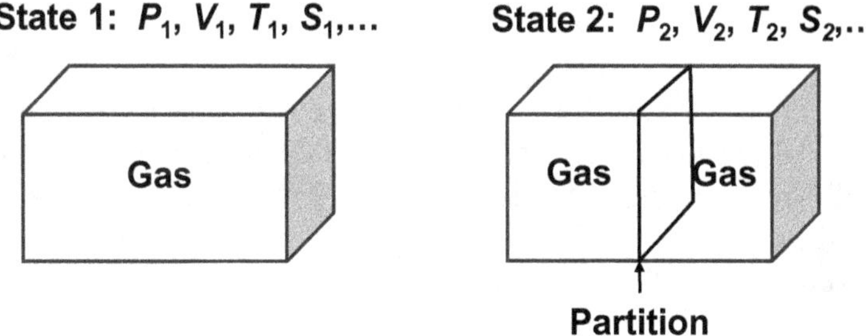

Figure 14.3. Extensive and Intensive Properties.

Processes can be either reversible or irreversible. A process is reversible if the system changes from an initial state to a final state by passing through a continuous sequence of equilibrium states. An equilibrium state is the state of a closed system that is obtained when the state variables are no longer changing with time. The state of the system can be made to reverse its path at any stage of the reversible process and return to its initial state. If the intermediate states are not equilibrium states, the process is irreversible, and it is not possible to return to the initial state from the final state by reversing the path of the process. All real processes are irreversible to some extent. In some cases, it is possible to approximate an irreversible process as a reversible process.

Let us now consider the cell of gas shown as State 1 in Figure 14.3. Suppose we subdivide the cell of gas into two halves by inserting the vertical partition shown in State 2. The subdivision is reversible if the partition has the same temperature as the gas in the cell and has negligible volume. Since any real partition will have a finite volume, we have chosen a partition that is small relative to the volume of the cell so that we can neglect the size of the partition and work with an approximately reversible process. If the gas was initially in an equilibrium state, the gas in each half of the cell should have the same pressure and temperature after inserting the partition as it did before the partition was inserted. The mass and volume in each half of the cell will be one-half of the original mass and volume, but their ratio, the density, is unchanged. Mass and volume are examples of extensive properties. An extensive property is a property of the system that depends on the amount of material. Density, temperature, and pressure are examples of intensive properties. An intensive property is a property of the system that is independent of the amount of material.

Entropy

Energy is often lost as waste heat in a thermodynamic process. Entropy is a measure of the amount of energy that is not available to do work. It is a physically measurable, extensive property. Entropy is a state variable and is labeled by '*S*' in Figure 14.3. German physicist Rudolf Clausius (1822–1888) coined the term 'entropy':

> **Q14.7**. "I propose to call the magnitude S the entropy of the body, from the Greek word τροπη, *transformation*. I have intentionally formed the word *entropy* so as to be as similar as possible to the word *energy*; the two magnitudes to be denoted by these words are so nearly allied in their physical meanings, that a certain similarity in designation appears to be desirable." (Clausius, 1867, p. 354; see also Daub, 1970, p. 321 footnote #2)

The change in entropy for a reversible process is equal to heat transfer from one state to another divided by the absolute temperature of the process.

Figure 14.4 shows the expansion of a gas from volume V_1 in State 1 to the larger volume V_2 in State 2. Mass M_1 in State 1 does not change; therefore, mass M_2 in State 2 equals mass M_1 in State 1. The increase in volume as a result of expansion results in an increase in entropy so that entropy S_2 in State 2 is larger than entropy S_1 in State 1.

Expansion of space in the aftermath of the Big Bang corresponds to an increase in volume. Combining the increasing volume of space with the decreasing temperature of the universe following the Big Bang yields an increase in entropy as cosmic time increases. This suggests that entropy was lowest at the moment of the Big Bang and has been increasing ever since.

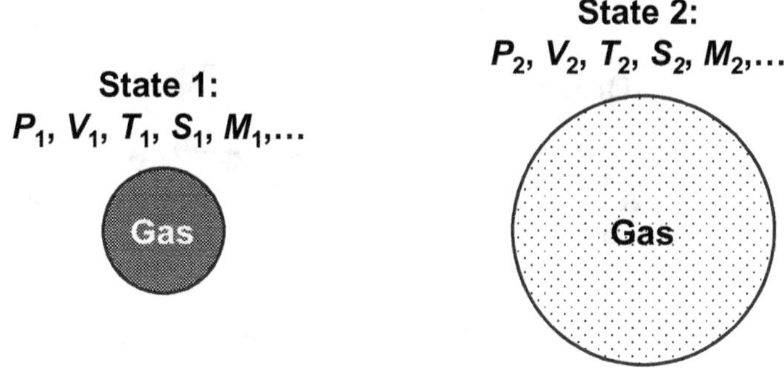

Figure 14.4. Expansion.

The laws of thermodynamics

The laws of thermodynamics are empirically based. For example, consider three systems *A*, *B*, and *C* with temperatures T_A, T_B, and T_C. If two of the systems are in thermodynamic equilibrium with the third system, then they must be in thermodynamic equilibrium with each other. This observation is the zeroth law of thermodynamics.

The first law of thermodynamics recognizes that heat is a form of energy. It is an expression of the conservation of nonrelativistic energy.

The second law of thermodynamics says that a spontaneously occurring process leads to an increase in entropy (*S*). A spontaneous process occurs without any outside intervention, otherwise it is a nonspontaneous process. An example of a spontaneous process is the flow of heat from a hot region to a cold region. This example does not require outside intervention and entropy increases. On the other hand, heat cannot transfer energy spontaneously from a colder region to a hotter region because the entropy of the overall system would decrease.

A system is in its ground state at zero absolute temperature. The third law of thermodynamics says that the entropy of the system approaches a constant value as the absolute temperature of the system approaches zero.

14.2 Entropy and Statistical Mechanics

Entropy was defined above as a measure of the amount of energy that is not available to do work. Another definition states that entropy is a

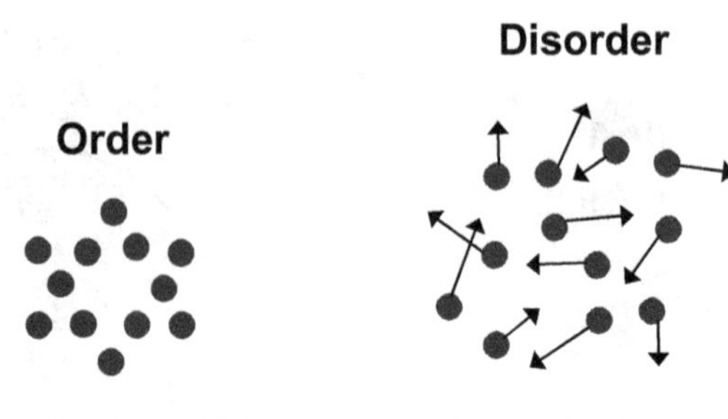

Figure 14.5. Order and Disorder.

measure of the disorder of a system. Figure 14.5 displays an ordered state, State 1, and a disordered state, State 2. The arrows indicate the direction and magnitude of motion. Entropy is larger in the more disordered state (water) than it is in the ordered state (snowflake).

The process of creating snowflakes from water or water vapor is a common occurrence. This process seems to increase order rather than disorder. The second law of thermodynamics prohibits the overall decrease in entropy, although a local decrease in entropy is allowed. The formation of astronomical systems such as stars and galaxies, the existence of living organisms, and the development of social structures are examples of systems that seem to be the result of increasing order and a decrease in entropy. It appears that entropy can decrease locally as long as it increases universally.

Entropy is related to the number of states that are accessible to a system. For example, suppose we are interested in counting the number of accessible states in a system of two 6-sided dice. If we roll the dice and sum the number of dots showing on the top faces of the dice, the accessible states range from a value of 2 to a value of 12. Any other values, such as 1 or 13, are not possible and therefore not accessible.

Figure 14.6 illustrates the likelihood of achieving two different states in a system of gas contained in an enclosure at a constant temperature and constant volume. State 1 in the figure is the most likely (probable) distribution of gas molecules because of random motion. State 2 in the figure with gas molecules clustering in a corner of the enclosure is unlikely (improbable).

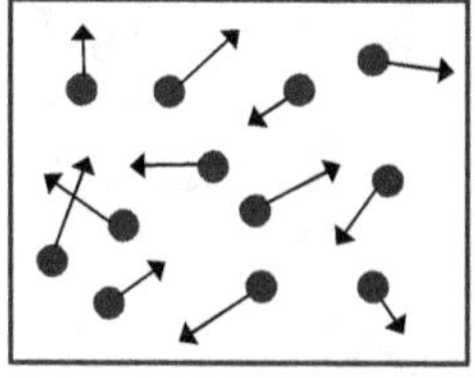
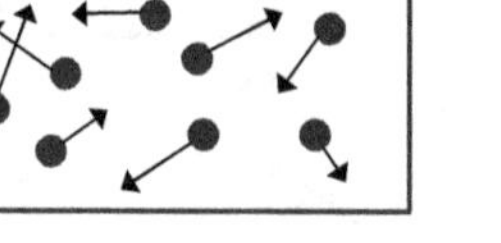
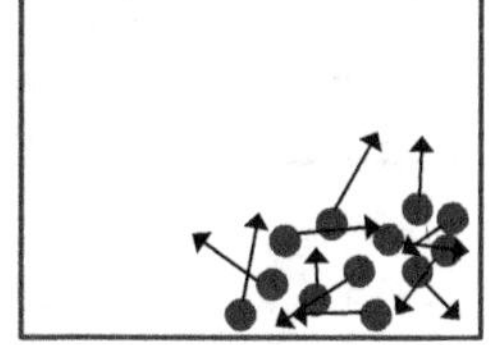

Figure 14.6. Likely and Unlikely States.

Physicists were divided at the end of the 19th century over the question of the reality of atoms. One faction believed that atoms were a useful mathematical construct but not real, while another faction believed that atoms were real and supported by the success of theories like the kinetic theory (Rooney, 2015, p. 33). Austrian physicist Ludwig Boltzmann (1844–1906) was a proponent of the kinetic theory and believed in the reality of atoms. Boltzmann analyzed the mechanical properties of collections of atoms using probability and statistics to determine the properties of macroscopic matter. He related the statistical behavior of atoms to macroscopically observable properties. His methods are known today as statistical mechanics.

An illustration of the statistical mechanical view is presented in Figure 14.7. Two insulated enclosures containing gas in States 1 and 2 are shown in the figure. The gas in State 1 and the gas in State 2 are in thermal equilibrium, that is, they are in the most probable (or likely) configuration within their respective enclosures. The most probable configuration is the configuration that has the maximum number of accessible states. If we allow the gases to mix as shown in the figure, we obtain a new equilibrium state corresponding to the configuration of combined gases with a new maximum number of accessible states.

Boltzmann recognized that entropy was a measure of the disorder of a system. On a more technical level, he showed that entropy was proportional to the logarithm of the number of accessible states. The proportionality constant is now called Boltzmann's constant and is considered a fundamental constant of nature.

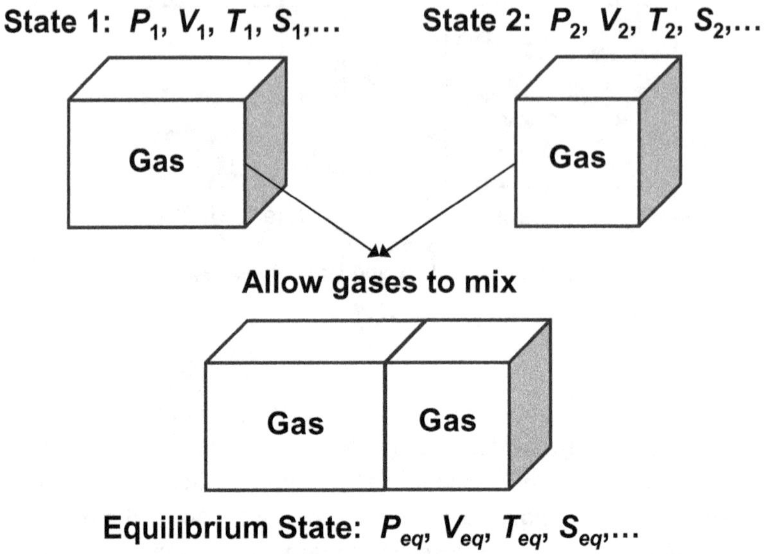

State 1: $P_1, V_1, T_1, S_1, \ldots$ **State 2:** $P_2, V_2, T_2, S_2, \ldots$

Gas Gas

Allow gases to mix

Gas Gas

Equilibrium State: $P_{eq}, V_{eq}, T_{eq}, S_{eq}, \ldots$

Figure 14.7. Equilibrium.

14.3 Time's Arrow

Arthur Eddington was the first to suggest a relationship between entropy and time's arrow:

> **Q14.8**. "So far as physics is concerned time's arrow is a property of entropy alone." (Eddington, 1929, p. 80)

Eddington noticed that the only law of physics that had an arrow of time was the Second Law of Thermodynamics:

> **Q14.9**. "The direction of time's arrow could only be determined by that incongruous mixture of theology and statistics known as the second law of thermodynamics; or, to be more explicit, the direction of the arrow could be determined by statistical rules, but its significance as a governing fact 'making sense of the world' could only be deduced on teleological assumptions." (Eddington, 1929, p. 338–339)

Our interest here is time's arrow.[4]

Cosmic time can be considered a temporal parameter for ordering cosmological events. The result is a timeline of cosmological events that correlates cosmic time with universal expansion and its associated declining temperature. As far as we know, cosmic time cannot be reversed. Cosmic time behaves like time's arrow.

The direction of time's arrow

Italian physicist Carlo Rovelli (1956–) said that "the entire difference between past and future may be attributed solely to the fact that the entropy of the world was low in the past" (Rovelli, 2015, p. 143). American physicist Richard A. Muller (1944–) explained the direction of time's arrow by observing that

> **Q14.10**. "time moves forward because our current state is so highly improbable. We have big concentrations of mass, lots of empty space, nonuniform temperatures… heat can flow, objects can break, and mass can disperse through empty space." (Muller, 2016, p. 123)

The expansion of the universe in the aftermath of the Big Bang provided space and matter that "was no longer at the maximum entropy possible for the new larger space" (Muller, 2016, p. 125). According to this view and the second law of thermodynamics, the current entropy of the expanding universe must be relatively low, and it is reasonable to expect it to increase.

Muller doubted Arthur Eddington's explanation that time's arrow was due to the increase in entropy. He suggested that the discovery of the Higgs boson cast suspicion on the claim that a causal relationship existed between entropy and time. The Higgs boson did not appear in the early universe until the universe cooled enough to allow spontaneous symmetry breaking. Before that event, the universe was apparently expanding with massless particles. This view depends on the validity of quantum field theory and the assumption that particles acquire mass by interacting with the Higgs boson. Muller theorized

> **Q14.11**. "that in the early universe, the entropy of all the matter was in massless, thermalized particles, so it [entropy] wasn't increasing. If time's arrow were truly being driven by the increase in entropy, there

> would have been no arrow. Time should have stopped… With stopped time, the expansion [of the universe] would have stopped." (Muller, 2016, p. 170)

The lack of causality between entropy and time in the early universe implies that Eddington's arrow of time idea did not apply during the early universe and cast doubt on its validity.

Lemaître's frame and the moment NOW

Another arrow of time is illustrated by Lemaître's frame. American physicist Richard A. Muller identified a reference frame he called Lemaître's frame in which the cosmological principle holds. An observer in Lemaître's frame sees the universe expanding because of the expansion of space between galaxies. All of the galaxies in Lemaître's frame are almost at rest. According to Muller, all of the galaxies in Lemaître's frame

> **Q14.12**. "have experienced the same time since the Big Bang, and all the clocks will read the same. That means they will all experience 'now' simultaneously." (Muller, 2016, p. 293)

Muller argued that the flow of time is set by the Big Bang rather than entropy. He said that the future does not exist because it is still being created. To Muller, the moment 'now' is at the boundary between past and future. He observed that

> **Q14.13**. "the time axis for a true spacetime diagram does not extend to infinity. Time stops at 'now'." (Muller, 2016, p. 294)

We focus on the present, says Muller, because "'Now' is the only moment when we can exercise influence" (Muller, 2016, p. 294). Muller recognized the Lemaître frame as "the only frame in which all the 'nows' across the universe are being created simultaneously" (Muller, 2016, p. 294).

Muller's point of view differs from the block universe model introduced in Chapter 6. Proponents of the block universe believe that all of spacetime already exists and is unchanging. The block universe is an example of eternalism.

Is time's arrow real?

Eddington's view of entropy and time's arrow is being challenged by modern researchers. As an example, we considered some of Muller's concerns about Eddington's view above. Here we outline Rovelli's hypothesis that time's arrow is not real; it is an illusion.

Rovelli said that the universe is a set of events that are not ordered in time. The events appear to imply relationships between physical variables in a relatively small part of the universe. Each part of the universe is a subsystem of the universe that interacts with a part of the physical variables to determine the state of that particular subsystem. The states accessible to each subsystem are a subset of the states that are accessible to the universe. Therefore, the universe appears to be in a high-entropy configuration with respect to each subsystem because there are many more states accessible to the universe. Rovelli introduced the thermal time hypothesis that

> **Q14.14**. "there is a flow associated with high-entropy configurations, and the parameter of this flow is thermal time." (Rovelli, 2017, pp. 155–156)

Entropy in a typical subsystem is relatively high along the flow of thermal time. However, there will be a few subsystems in which fluctuations of the entropy result in a few subsystems having low entropy relative to the average entropy of the universe. For these subsystems, entropy can increase. From this perspective, the growth in entropy

> **Q14.15**. "is what we experience as the flowing of time. What is special is not the state of the early universe: it is the [subsystem] to which we belong." (Rovelli, 2017, pp. 155–156)

Rovelli recognized the possibility of an alternative hypothesis:

> **Q14.16**. "The alternative is to accept as a given of observation the fact that entropy was low at the beginning of the universe." (Rovelli, 2017, pp. 156–157)

He suggested that low entropy in the past could have left detectable evidence in the present:

> **Q14.17**. "It is the presence of abundant traces of the past that produces the familiar sensation that the past is determined. The absence of any analogous traces of the future produces the sensation that the future is open." (Rovelli, 2017, p. 167)

# # #

We seem to sense an arrow of time even though some modern researchers are suggesting that our sense of motion through time is illusory. The reality of time is being reconsidered by researchers attempting to develop a theory of quantum gravity, which is the subject of the next chapter.

Endnotes

1. Sources that discuss heat and thermodynamics include d'Abro (1951), Feynman *et al.* (1963), Reif (1965), Muller (2016), Rovelli (2017), Carroll (2019), Çengel *et al.* (2019), and Urone *et al.* (2020).
2. Wolff (1965, 1967) contains a series of readings and essays on historically significant achievements in the physical sciences.
3. Sources that discuss kinetic theory and statistical mechanics include Reif (1965), Feynman (1998), Uffink (2004), Muller (2016), Rovelli (2017), Carroll (2019), and Urone *et al.* (2020).
4. For a discussion of science and teleology, see Fanchi (2021) and references therein.

Chapter 15

Time: Death or Rebirth?

Science has developed a narrative of the origin and evolution of the physical universe that is a patchwork of relativistic and quantum concepts. In this chapter, we briefly describe early attempts to unify the modern description of four fundamental interactions and then introduce quantum gravity as a theoretical attempt to reconcile differences between general relativity and quantum field theory. One objective of quantum gravity research is to obtain spacetime and gravity as emergent properties of some aspect of quantum theory, such as the wave function of the universe. Some leading approaches to quantum gravity that seek to eliminate time as a fundamental variable are described here, followed by a discussion of research that argues for a rebirth of the importance of time.

15.1 Grand Unified Theory and a Theory of Everything

Four fundamental interactions were identified in Chapter 12: strong, electromagnetic, weak, and gravitational. The mainstream view of physics is that a theory can be formulated which will unify these four seemingly disparate interactions into a single type of interaction.[1] Many people have attempted to unify the fundamental interactions. Einstein, for example, tried to develop a theory that unified the gravitational and electromagnetic interactions. He was one among many other researchers who have been unable to unify all four interactions. The status of the unification of fundamental interactions is summarized here.

The 1979 Nobel Prize in Physics was awarded to Steven Weinberg (1933–2021), Abdus Salam (1926–1996), and Sheldon Lee Glashow (1932–) for their work in developing the electroweak theory, that is, a unified theory of the weak and electromagnetic interactions. The combination of interactions is now known as the electroweak interaction and spurred efforts to unify the remaining two interactions.

Theories that attempt to combine strong, electromagnetic, and weak interactions are called grand unified theories. Glashow challenged the term when he observed that the gravitational interaction remains elusive:

> **Q15.1.** "…a synthesis of weak, strong, and electromagnetic interactions has been called a 'grand unified theory', but a theory is neither grand nor unified unless it includes a description of gravitational phenomena. We are still far from Einstein's truly grand design." (Glashow, 1980, p. 539)

A theory that combines all four interactions is called a theory of everything (TOE). TOE should include a grand unified theory and a theory of gravity as special cases, as illustrated in Figure 15.1. The grand unified theory (GUT) combines quantum field theories of the strong and electroweak interactions. A quantum field theory known as quantum chromodynamics (QCD) describes the strong interaction. It uses gluons to mediate the interaction between quarks. The strong interaction is associated with a quantum property assigned to quarks and gluons called the

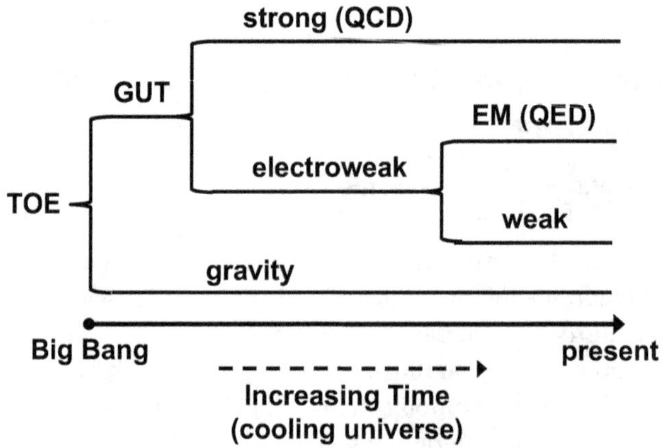

Figure 15.1. Theory of Everything.

color charge. The electroweak interaction is a quantum field theory of the electromagnetic (EM) and weak interactions. Quantum electrodynamics is a quantum field theory of the EM interaction.

Figure 15.1 is a sketch of a timeline that ranges from the moment of the 'Big Bang' to the present. It highlights changes to the dominant interactions in the universe at different points in time. The dominant interactions and corresponding theories change as the temperature of the universe decreases. The decrease in temperature associated with cooling allows transitions to occur from one dominant interaction to other interactions. The figure shows TOE transitioning to the gravitational interaction and GUT. The strong and electroweak interactions separate at the next transition. The electroweak interaction separates into EM and weak interactions as the universe continues to cool.

Transitions shown in Figure 15.1 are now thought to occur when there is a change in the symmetry of a quantum state. A state that is invariant or unchanging with respect to a set of transformations is considered symmetric with respect to that set of transformations. For example, a state is symmetric with respect to the Lorentz transformation of spacetime if the state is invariant with respect to the Lorentz transformation. Physicist Philip Anderson pointed out that the state "must always have the same symmetry as the laws of motion which govern it" (Anderson, 1972, p. 394). In cosmology, the symmetry of the initial state of the universe changes as a result of a transition between dominant interactions. The resulting change is known as symmetry breaking.

The Core Theory

Physicist Sean Carroll called a preliminary TOE that combines particle physics and general relativity a Core Theory. It is obtained by quantizing classical field theories, including general relativity, into a quantum field theory. The resulting quantum field combines the electroweak theory, QCD, and general relativity. The building blocks of nature are fields that pervade space, and particles are vibrations in the fields as discussed in Chapter 12. The Core Theory is not comprehensive, since

> **Q15.2.** "the Core Theory accurately describes not only particle physics but also gravity, as long as the strength of the gravitational field doesn't grow too large." (Carroll, 2019, p. 230)

Systems with strong or extreme gravitational fields, such as black holes or shortly after the Big Bang, are beyond the scope of the Core Theory.

15.2 The Many-Worlds Interpretation Revisited

A currently promising approach to a TOE is quantum gravity,[2] which encompasses theories that are attempting to restructure general relativity according to principles of quantum theory. Since there is no widely accepted theory of quantum gravity, we illustrate quantum gravity concepts using a theory of quantum gravity based on the Many-Worlds Interpretation (MWI). The MWI was introduced in Chapter 11. We apply MWI to Schroedinger's cat to further develop its basic concepts.

Schroedinger's cat's scenario is illustrated in Figure 11.7. A live cat is enclosed in a chamber with a bottle of cyanide, a radioactive atom, and an automated hammer. The hammer will break the bottle when the radioactive atom decays. We do not know when the atom will decay, but we know it will happen. Following physicist Sean Carroll's analysis (Carroll, 2019), we use square brackets [...] to refer to a classical state, and parentheses (...) to refer to a quantum state. The case of a classical cat being either alive or dead is expressed as

SC15.1. [cat] = [alive] or [cat] = [dead].

In this scenario, a quantum cat is in a superposition of a live cat and a dead cat (Figure 11.7):

SC15.2. (cat) = (alive + dead).

The quantum cat is then viewed by a classical observer abbreviated as [obs], thus

SC15.3. (cat) [obs] = (alive + dead) [obs].

In the Copenhagen interpretation, the cat is in a superposition of alive and dead. The wave function collapses to either the cat is [alive] or the cat is [dead] when the classical observer makes an observation. The result of the observation is the collapse of the wave function shown in Figure 11.9. The resulting state is either

SC15.4A. (cat) [obs] = (alive) [obs saw alive cat],

or

SC15.4B. (cat) [obs] = (dead) [obs saw dead cat].

American physicist Hugh Everett III, founder of the MWI, hypothesized that

1. The wave function represents reality.
2. The wave function does not collapse; it always obeys a temporal evolution equation like the Schroedinger equation.

There is one universal wave function in the Everettian view, and all aspects of the universe should be considered quantum. Classical physics should emerge from quantum physics. The universal wave function is also known as the wave function of the universe. In general, matter is entangled in the universal wave function.

The Everettian interpretation of Schroedinger's cat differs from the more traditional Copenhagen interpretation outlined in steps SC15.1 through SC15.4. Here the cat and the observer are both treated as quantum states:

SC15.5. (cat) (obs) = (alive + dead) (obs).

Measurement occurs when two different parts of the universe interact. The measurement (or observation) is a process that entangles the state of the cat with the state of the observer:

SC15.6. (cat) (obs) = (alive, obs sees alive cat + dead, obs sees dead cat).

Do we notice that, as observers, we are in a superposition of states with two different outcomes? If we do not sense the entanglement, does the prediction match reality? On the other hand, Carroll related the exchange between physicist Bryce DeWitt and Hugh Everett. DeWitt wrote a letter to Everett

Q15.3. "complaining that the real world obviously didn't 'branch,' since we never experience such things. Everett replied with a reference to

Copernicus's similarly daring idea that the Earth moves around the sun, rather than vice versa: 'I can't resist asking: Do you feel the motion of the earth?' DeWitt had to admit that was a pretty good response." (Carroll, 2019, p. 126)

The next step in the process is to recognize that we need to include the rest of the universe in the wave function. The part of the universe that is not the system of interest or the measuring apparatus is called the environment. We denote the environment as the quantum state (env), so that step SC15.5 becomes

SC15.7. (cat) (env) (obs) = (alive + dead) (env) (obs).

Carroll then describes the relationships between measurement, decoherence, branching, and the observer:

Q15.4. "A measurement is any interaction that causes a quantum system to become entangled with the environment, creating decoherence and a branching into separate worlds, and an observer is any system that brings such an interaction about." (Carroll, 2019, p. 122)

The cat becomes entangled with the environment when a measurement is made:

SC15.8.
(cat, env) (obs) = (alive, env with alive cat + dead, env with dead cat) (obs).

The interaction between the environment and the cat depends on whether the cat is alive (and breathing) or dead. The observer that makes the measurement (or observation) becomes entangled with the cat and the corresponding env:

SC15.9.
(cat, env, obs) = (alive, env with alive cat, obs sees alive cat + dead, env with dead cat, obs sees dead cat).

Decoherence splits the wave function (cat, env, obs) into branches that represent different outcomes (Figure 15.2). One outcome in the Schroedinger's cat's scenario is that the cat is alive and the other outcome

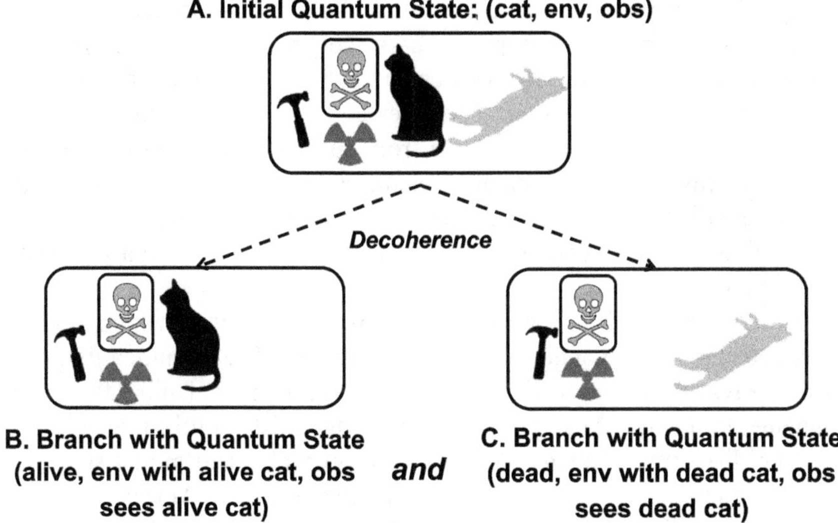

Figure 15.2. Sketch of the Many-Worlds Interpretation of Schroedinger's Cat.

is that the cat is dead. The different branches represent Everett's worlds in the Many-Worlds interpretation. The observer branches into

> **Q15.5**. "multiple copies along with the rest of the universe. After branching, each copy of the original observer finds themselves in a world with some particular measurement outcome." (Carroll, 2019, p. 119)

The observer does not sense the entanglement of states because a copy of each observer is now present in each branch. All possible measurement outcomes are real, but they exist in separate worlds along with a copy of each observer. The different branches are distinct and do not interact with one another.

Branching implies a direction of time. There were fewer branches of the wave function in the past. Similarly, entropy was lower in the past. Both branching and entropy have been increasing since the Big Bang. It is possible that branching is equivalent to an entropy-based arrow of time.

15.3 Quantum Gravity

Modern physicists typically start with classical theories and then apply a procedure that 'quantizes' the classical theory to develop a new quantum theory. There is no reason to believe that the quantum aspects of nature emerged from classical features such as the position of an object. It is more likely that classical features emerged from quantum aspects of an intrinsically quantum nature. Proponents of MWI are attempting to show how classical features emerge from such quantum concepts as the universal wave function, entanglement, decoherence, and branching. The expectation that classical features emerged from a quantum universe is a rationale for developing quantum gravity from a Many-Worlds perspective.

In the context of QFT, gravity occurs when a graviton, which has not been detected, is exchanged between two masses. The Feynman diagram of the gravitational interaction is shown in Figure 15.3. Unlike the other fundamental interactions, gravity continues to resist quantization.

The difficulty of quantizing gravity from a classical model suggests that we consider another approach. Let us first recall that American physicist John Archibald Wheeler (1911–2008) succinctly described the relationship between spacetime and matter: "Spacetime tells matter how to move; matter tells spacetime how to curve" (Wheeler, 2010, p. 235). We experience spacetime curvature as the force of gravity. Rather than directly quantizing spacetime and general relativity, can we show that classical physics emerges from quantum physics?

One possible clue comes from QFT: the universe is a system of interacting fields rather than particles, as discussed in Chapter 12. Fields fill all of space, and matter appears as vibrations of the fields. We can compare the particle view of empty space and the QFT view of vacuum by subdividing space into regions such as the hexagons shown in Figure 15.4. We then consider what happens in each region. In a world of particles, the space between particles is empty space. In QFT, empty space contains fields. The space between particle-like vibrations contains fields in their lowest energy states known as vacuum states.

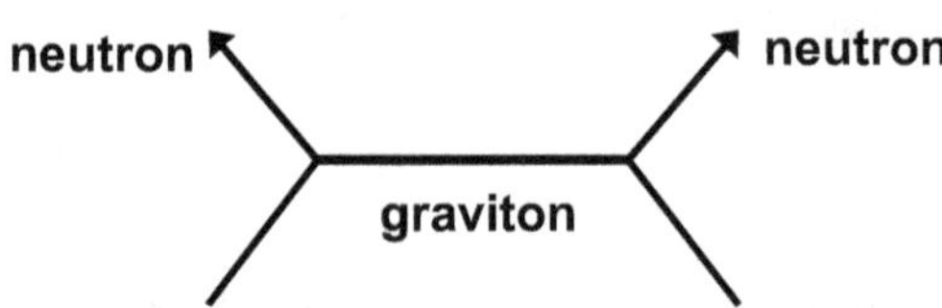

Figure 15.3. Feynman Diagram of the Gravitational Interaction.

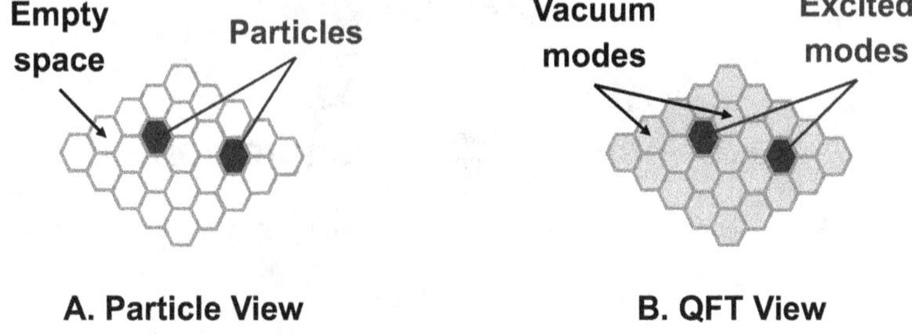

Figure 15.4. Empty Space and Vacuum.

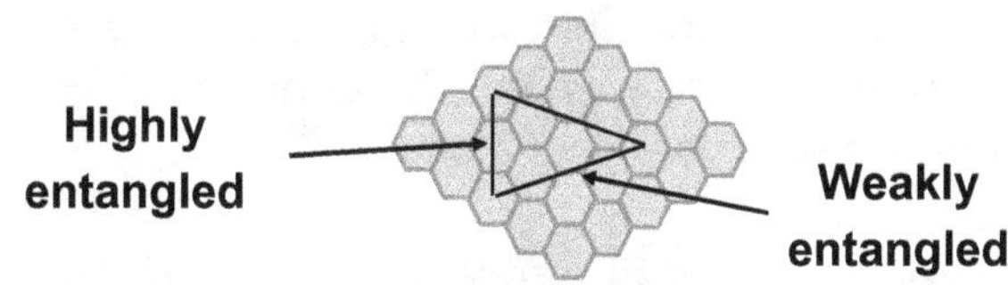

Figure 15.5. Does Spacetime Geometry Emerge from Entanglement?

Different regions of space are entangled with one another in QFT as shown in Figure 15.5. The degree of entanglement depends on the distance between regions: regions are highly entangled if they are close to one another. The degree of entanglement decreases with increasing distance between regions.

Another way to view the relationship between entanglement and distance is to express distance as a function of entanglement. The distance between regions is determined by the amount of entanglement between regions. The network of distance relationships defines the geometry of space. This provides a relationship between geometry and entanglement.

There is also a relationship between entanglement and mass–energy. The addition of particles to empty space increases mass–energy in the region. This breaks the entanglement of vibrating fields in the region with their surroundings. Entanglement in a region decreases when mass–energy is added to the region. Conversely, entanglement increases in a region when mass–energy is removed from the region.

Figure 15.6 illustrates connections between geometry, mass–energy, and entanglement. The goal of quantum gravity associated with the MWI is to use these relationships to show that general relativity can emerge from Everettian quantum theory.

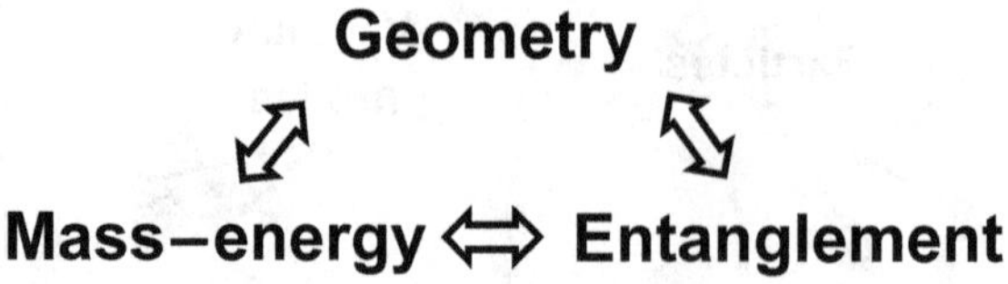

Geometry

⇗ ⇘

Mass–energy ⇔ Entanglement

Figure 15.6. Connections Associated with the Many-Worlds Interpretation.

Quantum gravitational effects appear to be important for understanding the microscopic structure of spacetime, early cosmology, and black holes. Since there is no accepted theory of quantum gravity, the term 'quantum gravity' is more descriptive of a research program than a particular theory. The Everettian approach to quantum gravity is one approach. Other approaches are possible:

> **Q15.6.** "In most, though not all, theories of quantum gravity, the gravitational field itself is also quantized. Since the contemporary theory of gravity, general relativity, describes gravitation as the curvature of spacetime by matter and energy, a quantization of gravity seemingly implies some sort of quantization of spacetime geometry: quantum spacetime." (Weinstein and Rickles, 2019)

Quantum spacetime in this context would have a microscopic granular structure. Quantum gravity is expected to provide a description of the microstructure of spacetime at a length scale known as the Planck scale ℓ_{Planck}, which is on the order of $\ell_{\text{Planck}} \sim 10^{-35}$ m.

The problem of time

The problem of time is a conceptual conflict between general relativity and quantum theory. It highlights the difficulties of describing the evolution in time of a quantum state or observable in canonical quantum gravity. Canonical quantum gravity is a program that treats spacetime as a field and attempts to quantize, or discretize, the field by separating spacetime into a stack of three-dimensional slices. A spatial geometry is defined on the slices of spacetime that include both time and space.

The preferred procedure for quantizing the classical theory in canonical quantum gravity is to impose constraints after quantization.

The constraint responsible for evolution in time in the classical theory is called the super-Hamiltonian. When it is quantized, its counterpart in the quantized theory seems

> **Q15.7**. "to indicate that the true physical states of the system do not evolve at all: there is no [time]. Trying to understand how, and in what sense, the quantum theory describes the time-evolution of something, be it states or observables, is the essence of the problem of time." (Weinstein and Rickles, 2019, Section 3.2.2)

The problem of time remains unresolved. For example, is time a real, fundamental physical property or is time an emergent property? We also note that physical laws at the microscopic scale do not seem to require time to flow in a single direction, yet our sense of time seems to flow into the future. Is the apparent asymmetry of time an emergent property of entropy and the second law of thermodynamics, or does it imply an alternative explanation?

15.4 The Block Universe and the Rebirth of Time

The block universe introduced in Chapter 6 is a universe with a history that is considered a single block of spacetime. The worldline of events in the block universe is always present and unchanging:

> **Q15.8**. "The past and future are reduced to events that exist together in the totality of a timeless, eternal block of spacetime." (Frank, 2011, p. 138)

The three-dimensional reality we perceive at each instant of time can be considered a slice of a four-dimensional spacetime.

Physicist Lee Smolin challenged the block universe view of time. Smolin said that the "world is presented to us as a series of moments" (Smolin, 2013, p. 92). We do not seem to have a choice about which moment we occupy. Furthermore, our experience of *Now* is not the same as our experience of the past as memory or the future as anticipation. The only moment when we seem to be able to make a choice is the moment *Now*. From a block universe perspective, the ability to make a choice is an illusion in a deterministic universe.

Smolin postulated that

> **Q15.9**. "to make sense of the picture of the universe that cosmological observations are bringing to us, we must embrace the reality of time in a new way. This is what I mean by the rebirth of time." (Smolin, 2013, p. xii)

He went on to say that he now believes

> **Q15.10**. "the deepest secret of the universe is that its essence rests in how it unfolds moment by moment in time." (Smolin, 2013, p. xiii)

According to Smolin,

> **Q15.11**. "the only meaningful way to define a quantity like time is to stipulate how to measure it. If you want to talk about time, you must describe what a clock is...and how the clock works. When you're approaching science operationally, you ask...what an observer could observe." (Smolin, 2013, p. 56)

The operational view of time is used to introduce historical time in Chapter 16.

# # #

Quantum gravity refers to any theory that is applicable to systems where gravity and quantum effects must be considered.

> **Q15.12**. "One should however remember that history teaches us that often a large majority of scientists has been working on the wrong tentative theory, with only a few working on the theory that ended up to be correct: scientific truth is judged by experimental corroboration, not by a democratic counting of votes!" (Rovelli, 2008)

In the preceding chapters, we considered concepts of time ranging from night and day to absolute time to relative time to the arrow of time to no time at all. Physicist Carlo Rovelli has adopted a relational view of time. He said that the fundamental theory of the world

> **Q15.13**. "does not need a time variable: it needs to tell us only how the things that we see in the world vary with respect to each other." (Rovelli, 2017, p. 119)

Rather than eliminating time as a fundamental variable, we argue that we need two types of time in the next chapter.

Endnotes

1. Sources that discuss grand unified theory (GUT) and the theory of everything (TOE) include Cooper and West (1988), Coughlan and Dodd (1991), Kaku (1993), Huang (2007), Hawking (2010), Carroll (2019), Ecker (2019), and Odenwald (2019).
2. Sources that discuss quantum gravity include Muller (2016), Rovelli (2008, 2017), Weinstein and Rickles (2019), and Carroll (2019).

Chapter 16

Historical Time

Time appears in many physical theories, including Newtonian and Einsteinian mechanics, Maxwell's electrodynamics, and quantum theory. Does time have the same meaning in all of these contexts? A basic flaw in understanding time stems from treating time as if it was the same physical quantity in every context in which it is used. An analogous situation occurred with the concept of mass. With the advent of Einsteinian relativity, it was legitimate to ask if inertial mass was the same as gravitational mass? Only by assuming the masses could be different was a deeper insight into the concept of mass attained by the scientific community.

When a similar treatment is accorded time, it is found that the concept of time is really three concepts with physically distinguishable characteristics. The three times are referred to here as Newtonian time, Einsteinian time, and historical time. Newtonian time is the usual nonrelativistic time of Newton and his predecessors. Einsteinian time is the time used in Einstein's theories of relativity and quantum field theories (QFT). Historical time is an invariant, monotonically increasing temporal parameter of parametrized relativistic quantum theory (pRQT).

The choice of times in physical theories is a choice between two hypotheses. The best way to understand these times is to compare the hypotheses and then develop an operational understanding, that is, an ability to measure the three times. We have already discussed Newtonian and Einsteinian time. Here we focus on historical time, an invariant evolution parameter.

16.1 Temporal Hypotheses

The relative merits of two temporal hypotheses were being evaluated during the first half of the 20th century. Einstein adopted what we refer to here as Hypothesis I, namely, the hypothesis that Newton's concept of absolute time must be replaced by the view that time was a component of a spacetime four-vector. The adoption of Hypothesis I led to the paradox that a single time, Einsteinian time, must satisfy two temporal concepts: time is a coordinate, and time is an arrow. Did scientists 'throw the baby out with the bath water' as illustrated in Figure 16.1?

Two key components of nonrelativistic quantum theory had to be modified to construct a relativistic quantum theory. A wave equation describing the evolution of the wave function had to be derived, and the equation that expresses the absolute square of the wave function as a probability distribution in space had to be extended to spacetime. The latter equation is known as a normalization condition.

German physicist Max Born introduced the normalization condition in the mid-1920s and described his Nobel prize-winning work on the statistical interpretation of quantum mechanics in 1954 (Born, 1954). He interpreted the wave function in probabilistic terms for nonrelativistic systems. According to Born's nonrelativistic normalization condition, once a massive particle is observed somewhere in the universe at some point in time, it should be possible to find the particle again at other times, even if we have to search the entire universe. At the time, scientists had not yet discovered unstable particles, and Born's nonrelativistic normalization condition was consistent with the experiment.

Figure 16.1. Discarding Newtonian Time.

Experimentally it is now known that most, if not all, particles are unstable. In other words, a particle with a particular identity at a certain point in spacetime may cease to have that identity at some later spacetime point. Pair annihilation and creation are examples of processes in which the identity of a particle is not maintained for all time.

A new theory is emerging which extends Born's concepts to include the entire realm of spacetime. It provides a probabilistic framework for describing unstable particles. The new theory is based on a temporal hypothesis that was first considered by Russian physicist Vladimir Aleksandrovich Fock (1898–1974) and Swiss physicist Ernst Carl Gerlach Stueckelberg (1905–1984) in the first half of the 20th century.[1] Fock and Stueckelberg introduced the idea that we should be working with two temporal variables. One temporal variable would be like Einsteinian time, and one temporal variable would be an independent invariant parameter. Fock and Stueckelberg treated the independent invariant parameter as a quantum-mechanical analog of proper time. Stueckelberg was the first to include a new normalization condition and a new interpretation of the probability function in addition to a parametrized wave equation. He was also the first to describe antimatter as matter traveling backward in time in the early 1940s. Shortly thereafter, American physicist Richard Feynman expressed a similar point of view that is now known as the Feynman–Stueckelberg interpretation (Boyarin and Land, 2008, p. 41).

Evolution parameter theories were part of the physics mainstream in the late 1940s and early 1950s when quantum field theory was being formulated. It helped guide such notable physicists as Richard Feynman (1918–1988) and Julian Schwinger (1918–1994) to develop quantum field theory. Researchers originally used the parameter as a mathematical construct with no physical significance. For example, Feynman talked about a "parameter u (somewhat analogous to proper time)" (Feynman, 1950, p. 453). He did not attribute a physical significance to the parameter.

A few groups made progress in developing evolution parameter theories in the intervening years, but their work was generally isolated. A resurgence of research in evolution parameter theories began in the 1970s with the publication of parametrized theories by American-Israeli physicist Lawrence P. Horwitz (1930–) and French-Belgian physicist Constantin Piron (1932–2012) in 1973, and independently by me and my Ph.D. advisor R. Eugene Collins (1925–2006) in 1978.[2]

Horwitz and Piron (1973) presented an axiomatic structure[3] for Stueckelberg's invariant evolution parameter they called τ:

> **Q16.1**. "We introduce a parameter τ to describe the evolution of [a many-particle] system. We call this parameter the historical time because it corresponds to the ordering relation determined by successive measuring processes in quantum theory or given by the laws of thermodynamics. One must be careful not to confuse the historical time τ..., which is an order relation, with the geometrical time t, one of the physical observables defining the state of a particle." (Horwitz and Piron, 1973, p. 317)

In a later paper, Arshansky, Horwitz, and Yeshiahu Lavie (1983) tried to understand τ in relation to measurements of an event in spacetime. They said that

> **Q16.2**. "measurements that we have available can, in practice, determine the occurrence of the event, but not the parameter τ which describes the evolution of its motion. Conventional laboratory clocks... measure the value of the observable t associated with the event, but we do not have available clocks which directly register the value of τ. It is important to emphasize that we do not rule out the existence of measurements which could, in principle, be performed at a given τ... We must be able to at least consistently conceive of performing measurements at a given τ." (Arshansky *et al.*, 1983, p. 1169)

These researchers did not have a procedure for measuring the parameter τ in the early 1980s. More than four decades after Horwitz and Piron resurrected Stueckelberg's invariant evolution parameter quantum theory, Horwitz and Arshansky (2018, Chapter 1) reminded the reader that

> **Q16.3**. "Horwitz and Piron (1973) postulated that, in order to construct many-body theories, there is just one universal τ, in complete accordance with Newton's postulate (1687) of a universal time." (Horwitz and Arshansky, 2018, Chapter 1)

Today, the axiomatic formulation introduced by physicists Horwitz and Piron (1973) is known as the SHP formulation, after Stueckelberg, Horwitz, and Piron.

The SHP formulation is one way to develop a pRQT. An alternative probabilistic formulation was developed by Fanchi and Collins. It was the relativistic extension of Collins' derivation of the nonrelativistic Schroedinger equation from probability theory (Collins, 1977). Shortly after I arrived at the University of Houston in 1975 as a Ph.D. student, I noticed that Collins' nonrelativistic derivation could be extended to a relativistic derivation of an equation like the Klein–Gordon equation. The Klein-Gordon-type equation turned out to be the Stueckelberg wave equation. This observation became the basis for my 1977 dissertation (Fanchi, 1977) and subsequent 1978 publications with Collins (see Endnote 2).

The modern formulation of pRQT can be expressed in several different ways. For example, both the axiomatic approach and the probabilistic approach led to Stueckelberg's parametrized relativistic wave equation and a normalization condition defined over a relativistic spacetime domain. Different formulations provide different perspectives and give us insight into the meaning of competing paradigms, such as pRQT and QFT.[4]

The unifying characteristic of parametrized relativistic quantum theories is the inclusion of an invariant evolution parameter to the usual spacetime coordinates.[5] The idea of adopting two temporal variables is formalized here as Hypothesis II, which says that relativistic quantum theory should include Einstein's coordinate time and an invariant evolution parameter that is the relativistic analog of Newtonian time.

16.2 Temporal Measurements

By the mid-1980s, the invariant evolution parameter τ had been called

> **Q16.4**. "proper time, an evolution parameter, and historical time. These many names reflect the lack of consensus among workers in the field on how the parameter should be viewed. As noted above, much of the confusion stems from ambiguity in the concept of time as used in different physical applications." (Fanchi, 1986, p. 1677)

A procedure was needed for measuring τ. One procedure for measuring the invariant evolution parameter and demonstrating its distinguishing characteristics is demonstrated by conducting a thought experiment (Fanchi, 1986, 2011). To avoid confusing the invariant evolution parameter with proper time, the symbol τ, which is often used to denote proper time, is replaced with the symbol s. The invariant evolution parameter s can be used to monitor the historical evolution of a system of events and is referred to here as historical time to emphasize this role.

Thought experiment for measuring a particle worldline

We can measure the particle worldline of a noninteracting particle in terms of spacetime coordinates by designing an experiment with clocks and detectors at events A and B as shown in Figure 16.2. Clocks and detectors measure the spacetime coordinates for each event as the particle traverses detectors A and B. We repeat the measurements a statistically significant number of times to obtain an ensemble of measurements.

The results of the spacetime measurements are shown in Figure 16.3. The figure presents the independent spacetime coordinates (x, t), and the statistical means of space and time $\langle x \rangle$, $\langle t \rangle$. Error bars display the uncertainty in the measurements. The primed frame F' on the right-hand side of the figure is moving with respect to the unprimed frame F. Each spacetime location is an event.

We can interpolate between spacetime events by introducing an independent, invariant evolution parameter s that lets us parametrize the

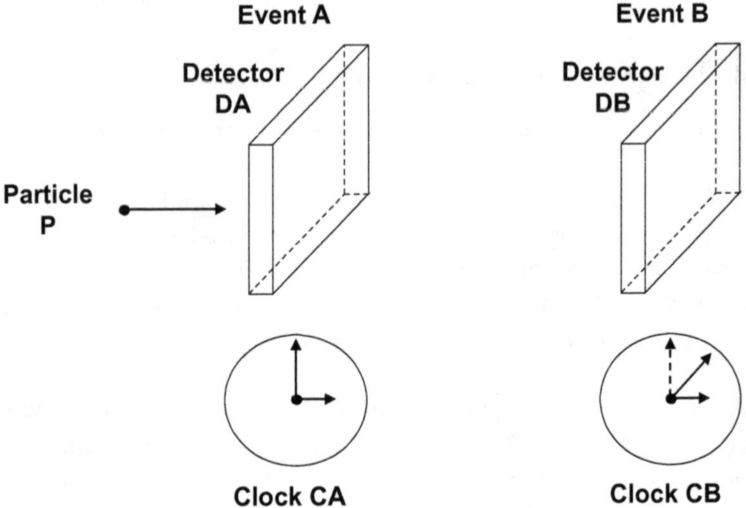

Figure 16.2. Measuring the Worldline of a Noninteracting Particle.

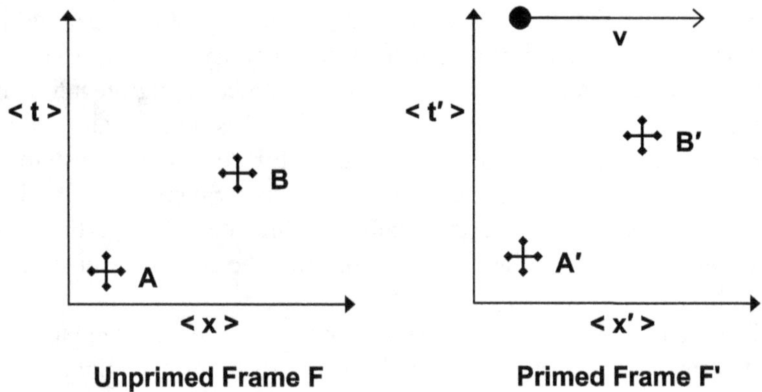

Figure 16.3. Plot Spacetime Observations.

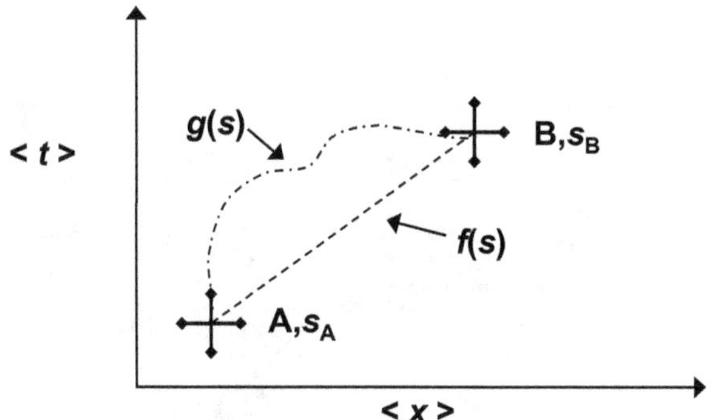

Figure 16.4. Parametrize Spacetime Observations.

spacetime observations. Figure 16.4 displays a parametrized particle worldline. The system is deterministic if we only consider a single worldline, such as $f(s)$. It is probabilistic if we consider many possible worldlines, such as $f(s)$ and $g(s)$.

The thought experiment demonstrates that it may be empirically useful to have an invariant evolution parameter.

The evolution parameter clock

A physically meaningful evolution parameter clock can be designed by recognizing that the invariant evolution parameter s is the relativistic analog of Newtonian time. The physical system should be separated into a subsystem for measuring parameter *s* and a subsystem that represents the experimental system of interest. There should be negligible interaction between the evolution parameter clock and the experimental system of interest. As an illustration of the procedure for designing a system that includes an evolution parameter clock and an experimental system, let us consider the behavior of two scalar particles: the interacting particle in the experimental system and the 's-clock' particle. The system is illustrated in Figure 16.5.

The 's-clock' particle trajectory may be used to replace the parameter *s* with familiar observables, such as spacetime coordinates. The noninteracting 's-clock' particle is introduced because it serves to independently define the evolution parameter *s*. The 's-clock' particle in this design propagates without interaction from a source to a detector. Its

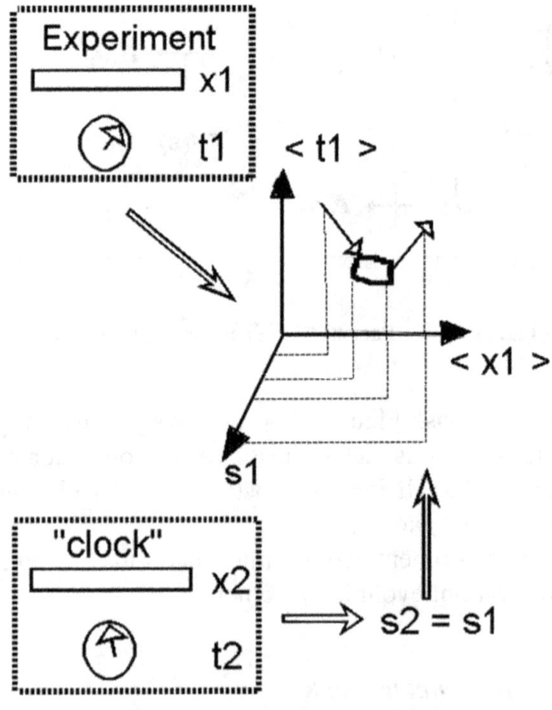

Figure 16.5. Evolution Parameter Clock.

trajectory is used to quantify evolution parameter s and parametrize the trajectory of the interacting particle in the experimental system. We measure the worldline of the 's-clock' particle using a rod with spatial coordinate $x2$ and clock with temporal coordinate $t2$. The value of the evolution parameter is then used to parametrize the experimental system. The assumption here is that the parameter of the evolution parameter clock can be equated to the parameter characterizing the experimental system. The same assumption is made in the nonrelativistic Schroedinger theory using Newtonian time (Fanchi, 1986) and is typical of the operational procedure we use to make "time" measurements in the nonrelativistic domain.

The "s-clock" analysis described above depends on the following assumptions:

a. The rest-frame clocks of the two particles are calibrated.
b. Particle 1 does not interact with particle 2 but may interact with other particles.
c. Particle 2 is a noninteracting particle.

Assumption (a) is a fundamental assumption that makes it possible to use a single evolution parameter in a many-body system. The remaining two assumptions serve to define the system of interest. Note that particle 1 and its interaction comprise the experimental system of interest. Particle 2 is an additional particle that serves as the evolution parameter clock.

An example of an evolution parameter clock is constructed using the two independent systems shown in Figure 16.6. The lower system in Figure 16.6 is an experimental system that experiences an interaction at the interaction point. The experimental system splits into a component that moves forward in time and a component that moves backward in time. The value of s is obtained from the upper system, which serves as an evolution parameter clock because it provides a monotonically increasing value of s.

The cosmic time described in Chapter 13 can be viewed as another example of an evolution parameter clock according to the measurement procedure outlined here. A clock moving with the Hubble flow provides a proper time that can be used as the historical time that is ordering events in a homogeneous, expanding universe.

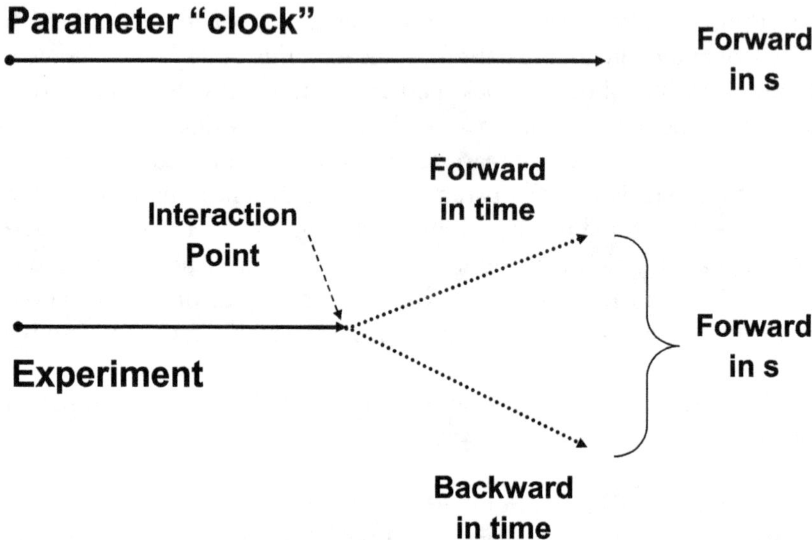

Figure 16.6. Evolution Parameter Clock.

16.3 Alternative Interpretations

When we recognize that Newtonian time is an evolution parameter in a Euclidean three-dimensional space and not a temporal coordinate in a Lorentzian four-dimensional space, we can gain insight into how to interpret temporal variables. Alternative interpretations of historical time have been provided by other researchers. American-Israeli physicist Martin Land [6] observed that a block universe

> **Q16.5**. "allows for no real changes, other than the inexorable motion of human consciousness through the predetermined script written into our trajectory through spacetime. In order to re-introduce dynamical evolution to physics, an additional measure of time must be defined." (Boyarin and Land, 2008, p. 29)

Land expressed a relational view that the configuration of objects in a physical system is characterized by separations in space and time at a given chronological moment:

> **Q16.6**. "This moment is determined by the flow of historical time, identical for all observers, and always moving forward. The change in spacetime configuration from one historical moment to another is called

evolution, and represents the physical influence of relationship on the interacting objects." (Boyarin and Land, 2008, p. 29)

Figure 16.7 illustrates a worldline between two spacetime events A and B. Events A and B represent the evolution of the spacetime (x^0, x^1) configuration in historical time s.

Land equated the parameter ordering the evolution of events along the worldline to historical time. According to this point of view, historical time

> **Q16.7**. "signifies the chronological evolution of events as they occur in historical sequence along the worldline, and is completely independent of the spacetime coordinates." (Boyarin and Land, 2008, p. 44)

Based on his interpretation of historical time, Land postulated a causal principle:

> **Q16.8**. "Events at one historical time cannot affect events at an earlier historical time... We will certainly not be able to change events that occurred at earlier historical times." (Boyarin and Land, 2008, p. 30)

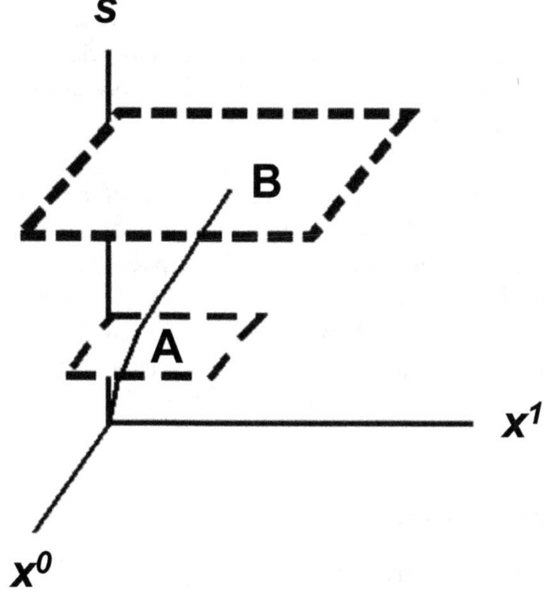

Figure 16.7. Historical Time.

The causal principle impacts our view of the grandfather paradox discussed in Chapter 6:

> **Q16.9**. "The so-called grandfather paradox – the objection to time travel on the grounds that one could change the past, and in the extreme case divert one's grandfather from the sequence of events that lead to one's birth — is absent in this formalism." (Boyarin and Land, 2008, p. 46)

Within the context of the SHP axiomatic formulation, Land realized that he could not change the chronological past, only the chronological future. Mueller arrived at a similar conclusion — but for different reasons, as discussed in Chapter 14 — when he sought to understand the meaning of *Now* (Mueller, 2016).

The range of possible interpretations of historical time is illustrated in the work of American mathematician Tepper Gill (1941–) and Slovenian physicist Matej Pavšič (1946–). Gill and his colleagues (2001) connected historical time with Isaac Newton's universal time. They showed that,

> **Q16.10**. "for any closed system of particles, there is a global inertial frame and unique (invariant) global proper-time (for each observer) from which to observe the system. This global clock is intrinsically related to the proper clocks of the individual particles and provides a unique definition of simultaneity for all events associated with the system. We suggest that this clock is the historical clock of Horwitz, Piron, and Fanchi." (Gill *et al.*, 2001, Abstract)

Pavšič (2001, 2020) investigated the role of parametrization in a variety of modern mathematical and physical contexts, such as QFT and brane space. He developed a Clifford space (C-space) formulation of a generalized Stueckelberg equation in 14-dimensional subspace of 16-dimensional C-space (Pavšič, 2020, Section 5.2). The evolution parameter analogous to historical time was related to two coordinates in C-space: the scalar and pseudoscalar coordinates.

16.4 Experimental Tests

One clue to finding experimental tests of Hypothesis II and pRQT is provided by nonrelativistic quantum mechanics. We know that an interaction potential that depends on Newtonian time in the nonrelativistic quantum theory can represent systems in which there are transitions between energy states. The resulting time-dependent theory is applicable to devices such as the laser. An analog in pRQT occurs when the interaction term is allowed to depend on the invariant evolution parameter *s*. In this case, pRQT can describe systems in which there are transitions between mass states. pRQT has been applied to mass state transitions in a variety of systems, such as neutral K-meson (K^0) decay, neutrino oscillations, and the creation and annihilation of tachyons with real mass.[7]

# # #

Time is one of the most accurately measured quantities in all of science, yet the literature — ranging from science to art — abounds with temporal paradoxes. Paradoxes in special relativity are often related to time travel and the time traveler interfering with the past worldline. The inclusion of the invariant evolution parameter *s* in pRQT leads to a dynamic universe where the time traveler cannot return to a specific past event on the worldline because spacetime is evolving with respect to *s* as shown in Figure 16.7. The presence of historical time in pRQT re-introduced dynamical evolution to the description of relativistic systems.

Endnotes

1. A review of theories with an evolution parameter prior to 1993 was presented by Fanchi (1993a). Pioneering papers by Fock (1937) and Stueckelberg (1941a, 1941b, 1942) are cited in the References.
2. The probabilistic formulation was first presented in Fanchi (1977), Collins and Fanchi (1978), and Fanchi and Collins (1978).
3. Note that Piron earned his Doctor of Science degree in 1963 under the direction of Ernst Stueckelberg and Josef-Maria Jauch at the University of Lausanne with a thesis on quantum logic. Horwitz earned his Doctor of Philosophy degree under Julian Schwinger at Harvard University in 1957 and worked at the University of Geneva from 1964 to 1966.
4. Many proponents of quantum field theory mistakenly believe that a quantum mechanical theory cannot represent the creation and annihilation of particles. Parametrized relativistic quantum theory is presented as a counterexample (Fanchi, 2021). It shows how particle stability can be represented by extending the normalization condition over spacetime and introducing a marginal probability density in time.

5. Sources that introduce parametrized theories based on Stueckelberg's work (1941a, 1941b, 1942) include Fanchi (1993b, 2011), Pavšič (2001, 2020), Horwitz (2015), Horwitz and Arshansky (2018), and Land and Horwitz (2020). Many other references are included in these sources.
6. Martin Land and Rafael Arshansky were doctoral students of Lawrence P. Horwitz.
7. Sources include neutral K-meson (K^0) decay (Fanchi, 2003), neutrino oscillations (Fanchi, 2019), and the creation and annihilation of tachyons with real mass (Fanchi, 2022).

Chapter 17

Epilogue — Resolving the Enigma of Time

People have tried to understand the role of time in nature for millennia. The history of time up to Newtonian time was reviewed in Part 1. Einsteinian time and its role in relativity were discussed in Part 2. Part 3 presented the application of Einsteinian time and spacetime to cosmology and the inflationary universe. Modern, and more speculative, theories were outlined in Part 4. Modern theories range from the block universe to quantum gravity to parametrized relativistic quantum theory.

The success of nonrelativistic quantum mechanics and special relativity motivated efforts to extend quantum concepts to the relativistic domain. The role of time was a key difference between Einsteinian and Newtonian views of classical theory. To Newton, time marched inexorably from the past to the future. Newtonian time was an arrow pointing to the future and representing an asymmetry in the flow of time.

Einstein's views required a physical equivalence between coordinate time and coordinate space. In this view, time should be a reversible coordinate in the same manner as space. Can time be both an arrow and a coordinate? How can time be an irreversible arrow linked to entropy and a reversible coordinate in the Einsteinian sense? Herein lies the crux of a temporal enigma, the mystery of time, and a clue to its resolution.

The temporal issues introduced in Chapter 1 are shown in Table 17.1. The last column in the table shows that the issues remain. For example, proponents of the block universe view spacetime as static and time as

Table 17.1. Temporal Issues.

What Is Time?	How Many Times?	Time Is	Still a Possibility?
Time is an illusion			Yes
Time is real	One time	Absolute or relative	Yes
	Two times	Absolute and relative	Yes
Time is emergent			Yes

illusory. On the other hand, proponents of quantum gravity are seeking to represent time as an emergent property.

Two hypotheses are possible if time is real. In the first hypothesis, Hypothesis I, one temporal variable is assumed to exist, and the temporal variable is Einsteinian time. The second hypothesis, Hypothesis II, relies on the existence of two temporal variables. One temporal variable is a coordinate time in the sense of Einstein, and the other temporal variable is an invariant evolution parameter in the sense of Newton. If Hypothesis I is correct, time must be both an irreversible arrow linked to entropy and a reversible coordinate in the Einsteinian sense.

The development of evolution parameter theories has been motivated in part by the concern that Hypothesis I is too restrictive. If Hypothesis II is correct, there are two temporal variables: Einsteinian time is a reversible coordinate in the sense of Einstein, and historical time is an irreversible, invariant evolution parameter.

It seems reasonable to conclude that many temporal issues remain and many theories are possible. In the noblest spirit of science, the theories must submit to experimental testing and be subject to validation.

Dates of Selected Characters

Name (Last, First)	Dates	Country of Birth
Alexander III of Macedon, aka Alexander the Great	356–323 BCE	Ancient Greece
Anaximander	ca. 610–546 BCE	Miletus, Ionia (now Turkey)
Aristarchus of Samos	ca. 310–230 BCE	Greece
Aristotle	ca. 384–322 BCE	Greece
Augustine of Hippo aka Saint Augustine	354–430	Algeria
Berkeley, George	1685–1753	England
Bohr, Niels	1885–1962	Denmark
Boltzmann, Ludwig	1844–1906	Austria
Born, Max	1882–1970	Germany
Brahe, Tycho	1546–1601	Scania, Denmark (now Norway)
de Broglie, Louis Victor	1892–1987	France
Childe, Verne Gordon	1892–1957	Australia
Clausius, Rudolf	1822–1888	German
Colding, Ludvig A.	1815–1888	Denmark

Name (Last, First)	Dates	Country of Birth
Collins, Royal Eugene	1925–2006	USA
Copernicus, Nicolaus	1473–1543	Poland
Dalton, John	1776–1844	England
Davisson, Clinton	1904–1991	USA
Democritus of Abdera	ca. 460–370 BCE	Thrace, northeast Greece
Descartes, René	1596–1650	France
Doppler, Christian Andreas	1803–1853	Austria
Eddington, Arthur	1882–1944	UK
Einstein, Albert	1879–1955	Germany
Euclid of Alexandria	ca. mid 400–mid 300 BCE	Greece
Eudoxus of Knidus	ca. 408–355 BCE	Asia Minor
Everett, Hugh III	1930–1982	USA
Faraday, Michael	1791–1867	England
Feynman, Richard P.	1918–1988	USA
FitzGerald, George Francis	1851–1901	Ireland
Fock, Vladimir Aleksandrovich	1898–1974	Russia
Friedmann, Alexander	1888–1925	Russia
Galilei, Galileo	1564–1642	Italy
Gamow, George	1904–1968	Odesa, Russian Empire (now Ukraine)
Gell-Mann, Murray	1929–2019	USA
Gerlach, Walther	1889–1979	Germany
Germer, Lester	1896–1971	USA
Gill, Tepper	1941–	USA
Glashow, Sheldon Lee	1932–	USA
Goudsmit, Samuel A.	1902–1978	Netherlands
Guth, Alan	1947–	USA

Name (Last, First)	Dates	Country of Birth
Halley, Edmund	1656–1742	England
Hawking, Stephen	1942–2018	UK
Heisenberg, Werner	1901–1976	Germany
Helmholtz, Hermann Ludwig von	1821–1894	Germany
Heraclitus	535–475 BCE	Ephesus, Ionia (now Turkey)
Hesiod	ca. 750–650 BCE	Ancient Greece
Higgs, Peter	1929–	UK
Hill, W.E. (William Ely)	1887–1962	USA
Hobbes, Thomas	1588–1679	England
Hooke, Robert	1635–1703	England
Horwitz, Lawrence Paul	1930–	USA
Hoyle, Fred	1915–2001	UK
Hubble, Edwin P.	1889–1953	USA
Hume, David	1711–1776	Scotland
Huygens, Christiaan	1629–1695	Netherlands
Joule, James	1818–1889	England
Joyce, James	1882–1941	Ireland
Kant, Immanuel	1724–1804	Germany
Kepler, Johannes	1571–1630	Germany
Kermer, Lester H.	1896–1971	USA
Kragh, Helge	1944–	Denmark
Kuhn, Thomas	1922–1996	USA
Lakatos, Imre	1922–1974	Hungary
Land, Martin	1953–	USA
Laplace, Pierre-Simon	1749–1827	France
Lavoisier, Antoine	1743–1794 (by guillotine)	France

Name (Last, First)	Dates	Country of Birth
Leavitt, Henrietta Swan	1868–1921	USA
Leibniz, Gottfried Wilhelm von	1646–1716	Germany
Lemaître, Georges	1894–1966	Belgium
Leucippus	ca. 5th century BCE	Greece
Lippershey, Hans	ca. 1570–1619	German-Dutch
Lorentz, Henrik Anton	1853–1928	Netherlands
Lowell, Percival	1855–1916	USA
Mach, Ernst	1838–1916	Austria
Maxwell, James Clerk	1831–1879	Scotland
Mayer, Julius	1814–1878	Germany
Mendeleev, Dmitri	1834–1907	Russia
Michelson, Albert A.	1852–1931	Poland (formerly Prussia)
Minkowski, Hermann	1864–1909	Poland (now Lithuania)
Morley, Edward W.	1838–1923	USA
Muller, Richard A.	1944–	USA
Newton, Isaac	1643–1727	England
Parmenides of Elea	ca. 515 BCE–?	Greek colony in Italy
Pauli, Wolfgang	1900–1958	Austria
Pavšič, Matej	1946–	Slovenia
Penzias, Arno A.	1933–	USA
Perrin, Jean-Baptiste	1870–1942	France
Piron, Constantin	1932–2012	France
Planck, Max	1858–1947	Germany
Plato from Athens	428–347 BCE	Greece
Popper, Karl	1902–1994	Austria
Powell, Cecil F.	1903–1969	UK

Name (Last, First)	Dates	Country of Birth
Priestley, Joseph	1733–1804	England
Ptolemy, Claudius	ca. 100–170	Egypt in Roman Empire
Pythagoras of Samos	ca. 570–490 BCE	Ionian Greece
Roland, Loránd, baron von Eötvös	1848–1919	Hungary
Rovelli, Carlo	1956–	Italy
Rubin, Vera	1928–2016	USA
Salam, Abdus	1926–1996	Pakistan
Schroedinger, Erwin	1887–1961	Germany
Schwinger, Julian	1918–1994	USA
de Sitter, Willem	1872–1934	Netherlands
Slipher, Vesto	1875–1969	USA
Socrates from Athens	ca. 470–399 BCE	Greece
Spinoza, Baruch	1632–1677	Portugal
Stern, Otto	1888–1969	Germany
Stueckelberg, Ernst Carl Gerlach	1905–1984	Switzerland
Thales	ca. 624–546 BCE	Miletus, Ionia (now Turkey)
Thompson, Benjamin aka Count Rumford	1753–1814	UK (born in USA)
Uhlenbeck, George E.	1900–1988	Netherlands
Ussher, James	1581–1656	Ireland
Weinberg, Steven	1933–2021	USA
Wheeler, John Archibald	1911–2008	USA
Wilson, Robert W.	1936–	USA
Yukawa, Hideki	1907–1981	Japan
Zweig, George	1937–	Russia
Zwicky, Fritz	1898–1974	Bulgaria

References

AIP Telescopes (2021). The First Telescopes. Center for the History of Physics. American Institute of Physics. Retrieved from https://history.aip.org/exhibits/cosmology/tools/tools-first-telescopes.htm on November 6, 2021.

Amadio, A.H. and A.J.P. Kenny (2021, March 2). Aristotle. Encyclopedia Britannica. Retrieved from https://www.britannica.com/biography/Aristotle on November 5, 2021.

Anderson, P.W. (1972). More Is Different. *Science*, *147*, 393–396.

Aristotle (350 BCE). *Physics. The Complete Works of Aristotle*, In Jonathon Barnes (ed.), The Revised Oxford Translation, Volume 1, 1991. Princeton University Press, Princeton, New Jersey.

Arshansky, R., L.P. Horwitz, and Y. Lavie (1983). Particles vs. Events: The Concatenated Structure of World Lines in Relativistic Quantum Mechanics. *Foundations of Physics*, *13*, 1167–1194.

Aughton, P. (2008). *The Story of Astronomy*. Quercus, London.

Augustine (ca. 398). *The Confessions of Saint Augustine*. Translated and edited by A.C. Outler in *Augustine: Confessions* (1955). Retrieved from https://faculty.georgetown.edu/jod/augustine/conf.pdf on November 11, 2021.

Bacciagaluppi, G. and A. Valentini (2009). *Quantum Theory at the Crossroads: Reconsidering the 1927 Solvay Conference*. Cambridge University Press, Cambridge, UK.

Baggott, J. (2013). *Farewell to Reality*. Pegasus Books, London.

Bambi, C. (2018). *Introduction to General Relativity*. Springer Nature, Singapore.

Barrett, J.A. and P. Byrne (2012). *The Everett Interpretation of Quantum Mechanics: Collected Works 1955–1980 with Commentary*. Princeton University Press, Princeton, New Jersey.

Barnes, J. (1991). *The Complete Works of Aristotle*, In Jonathon Barnes (ed.), The Revised Oxford Translation. Princeton University Press, Princeton, New Jersey.

Bennett, J., M. Donahue, N. Schneider, and M. Voit (2018). *The Essential Cosmic Perspective*, 8th Edition. Pearson, New York.

Bergstrom, L. and A. Goobar (1999). *Cosmology and Particle Astrophysics*. Wiley, New York.

Berkeley, G. (2012) *De Motu*. Included in *The Project Gutenberg eBook of the Works of George Berkeley*. Volume 1 of 4: Philosophical Works, 1705–21. First published by Clarendon Press, Oxford, UK, 1901. Retrieved from https://www.gutenberg.org/files/39746/39746-pdf.pdf on January 16, 2022.

Bernal, J.D. (1972). *The History of Classical Physics*. Published in 1999 by Barnes & Noble, New York.

Blandford, R. and N. Gehrels (1999). Revisiting the Black Hole. *Physics Today* (June), 40–46.

Bondi, H. and T. Gold (1948). The Steady-State Theory of the Expanding Universe. *MNRAS, 108*, 252–270.

Born, M. (1927). Quantenmechanik der Stoßvorgänge (Quantum Mechanics of Collision Processes). Translated by D.H. Delphenich. *Zeitschrift für Physik, 38*, 803–827.

Born, M. (1954). The Statistical Interpretation of Quantum Mechanics. Nobel Prize organization. Retrieved from https://www.nobelprize.org/uploads/2018/06/born-lecture.pdf on March 21, 2022.

Boyarin, J. and M. Land (2008). *Time and Human Language Now*. Prickly Paradigm Press, Chicago.

Brennan, R.P. (1997). *Heisenberg Probably Slept Here*. Wiley, New York.

Britannica, The Editors of Encyclopaedia (2012, March 6). Ionia. Encyclopedia Britannica. Retrieved from https://www.britannica.com/place/Ionia on November 4, 2021.

Britannica, The Editors of Encyclopaedia (2016, March 1). Urban Revolution. Encyclopedia Britannica. Retrieved from https://www.britannica.com/topic/urban-revolution on November 3, 2021.

Britannica, The Editors of Encyclopaedia (2018, January 17). Olbers' Paradox. Encyclopedia Britannica. Retrieved from https://www.britannica.com/print/article/426726 on January 16, 2022.

Britannica, The Editors of Encyclopaedia (2021, February 18). Mach's Principle. Encyclopedia Britannica. Retrieved from https://www.britannica.com/science/Machs-principle on February 4, 2022.

Britannica, The Editors of Encyclopaedia (2022, January 10). George Berkeley. Encyclopedia Britannica. Retrieved from https://www.britannica.com/biography/George-Berkeley on January 16, 2022.

Brouwer, W. (1980). Einstein and Lorentz: The Structure of a Scientific Revolution. *American Journal of Physics*, *48*, 425–431.

Bucher, M.A. and D.N. Spergel (1999, January). Inflation in a Low-Density Universe. *Scientific American, 280*, 63–69.

Burbridge, G., F. Hoyle, and J.V. Narlikar (1999). A Different Approach to Cosmology. *Physics Today, 52*, 38–44. A reply by A. Albrecht representing normal science is appended to the article, *Ibid.*, 44–46.

Byrne, P. (2007, December). The Many Worlds of Hugh Everett. *Scientific American, 297*(6), 72–79.

Carroll, S. (2012). *The Particle at the End of the Universe*. Penguin Group, New York.

Carroll, S. (2019). *Something Deeply Hidden*. Dutton-Penguin Random House, New York.

Çengel, Y., M. Boles, and M. Kanoglu (2019). *Thermodynamics: An Engineering Approach*, 9th Edition. McGraw-Hill, New York.

Clark, J. (2021). Vesto Slipher: Uncovering the Cosmos. Indiana History Blog, Indiana Historical Bureau of the Indiana State Library, September 1. Retrieved from https://blog.history.in.gov/vesto-slipher-uncovering-the-cosmos/ on February 5, 2021.

Clausius, R. (1865). On Several Convenient Forms of the Fundamental Equations of the Mechanical Theory of Heat. *The Mechanical Theory of Heat — with its Applications to the Steam Engine and to Physical Properties of Bodies.* Translation of this Ninth Memoir published by London: John van Voorst, 1 Paternoster Row. MDCCCLXVII, pp. 327–365; see also Clausius, R. (1865). Ueber verschiedene für die Anwendung bequeme Formen der Hauptgleichungen der mechanischen Wärmetheorie. In German. *Annalen der Physik, 125*, 353–400.

CMS Collaboration (2012). Observation of a new boson at a mass of 125 GeV with the CMS experiment at the LHC. *Physics Letters B, 716*, 30–61.

Collins, R.E. (1977). Quantum theory: A Hilbert space formalism for probability theory. *Foundations of Physics, 7*, 475–494.

Collins, R.E. and J.R. Fanchi (1978). Relativistic Quantum Mechanics: A Space-Time Formalism for Spin-Zero Particles. *Nuovo Cimento, 48A*, 314–326.

Conselice, C.J. (2007, February). The Universe's Invisible Hand. *Scientific American, 296*, 34–41.

Cooper, N.G. and G.B. West (eds.) (1988). *Particle Physics: A Los Alamos Primer*. Cambridge University Press, Cambridge, UK.

Copernicus, N. (1543). *De Revolutionibus Orbium Coelestium* (On the Revolutions of the Heavenly Spheres). Translation and Commentary by Edward Rosen. The Johns Hopkins University Press, Baltimore and London. Retrieved from http://www.webexhibits.org/calendars/year-text-Copernicus.html on November 6, 2021.

Coughlan, G.D. and J.E. Dodd (1991). *The Ideas of Particle Physics*, 2nd Edition. Cambridge University Press, Cambridge, UK.

Coward, F. (2009). *The Rise of Humans. Prehistoric Life.* DK Publishing, New York.

Cronk, G., T. de Marco, P. Dlugos, and P. Eckstein (2004). *Readings in Philosophy*, 2nd Edition. Hayden McNeil Publishing, Plymouth, Michigan.

Cushing, J.T. (1998). *Philosophical Concepts in Physics*. Cambridge University Press, Cambridge, UK.

d'Abro, A. (1951). *The Rise of the New Physics*. Dover, New York, 2 volumes.

Daub, E.E. (1970). Entropy and Dissipation. *Historical Studies in the Physical Sciences*, *2*, 321–354.

Davies, P. (2007). Universes Galore: Where Will it All End? In B. Carr (ed.), *Universe or Multiverse?* Cambridge University Press, Cambridge, UK, pp. 487–506. DOI:10.1017/CBO9781107050990.030.

Davisson, C. and L.H. Germer (1927). Diffraction of Electrons by a Crystal of Nickel. *Physical Review*, *30*, 705–740.

de Broglie, L. (1966). *Physics and Microphysics*. Grosset's Universal Library, Grosset and Dunlap, New York.

de Sitter, W. (1917a). On the relativity of inertia. Remarks concerning Einstein's latest hypothesis. Proceedings 19 paper 2 of the Royal Netherlands Academy of Arts and Sciences (KNAW), Amsterdam, pp. 1217–1225. Retrieved from https://sites.pitt.edu/~jdnorton/teaching/2590_Einstein_2015/pdfs/de%20Sitter%20Amst%20Acad%201917.pdf on February 3, 2022.

de Sitter, W. (1917b). On Einstein's Theory of Gravitation, and its Astronomical Consequences. Proceedings 19 paper 3 of the Royal Netherlands Academy of Arts and Sciences (KNAW), Amsterdam, pp. 3–28. Retrieved from https://articles.adsabs.harvard.edu/cgi-bin/nph-iarticle_query?1917MNRAS..78....3D&defaultprint=YES&filetype=.pdf on February 3, 2022; Translation in *Monthly Notices of the Royal Astronomical Society*, *91*(5), March 1931, 483–490. https://doi.org/10.1093/mnras/91.5.483.

DiSalle, R. (2020). Space and Time: Inertial Frames. The Stanford Encyclopedia of Philosophy (Winter 2020 Edition), In E.N. Zalta (ed.). Retrieved from https://plato.stanford.edu/archives/win2020/entries/spacetime-iframes/ on November 12, 2021.

Earman, J. and C. Glymour (1980). The Gravitational Red Shift as a Test of General Relativity: History and Analysis. *Studies in History and Philosophy of Science*, *11*, 175–214. History of Science Essay.

Ecker, G. (2019). *Particles, Fields, Quanta.* Springer Nature Switzerland AG, Cham, Switzerland.

Eddington, A. (1929). *The Nature of the Physical World.* MacMillan Company, New York and The University Press, Cambridge, UK. Based on 1927 Gifford Lectures. Third Impression. See also https://archive.org/details/in.ernet.dli.2015.23480, accessed September 2, 2022.

Edwards, P.P., R.G. Egdell, D. Fenske, and B. Yao (2020). The periodic law of the chemical elements: 'The new system of atomic weights which renders evident the analogies which exist between bodies' [1]. *Philosophical Transactions of the Royal Society A, 378*, 20190537. http://dx.doi.org/10.1098/rsta.2019.0537.

Einstein, A. (1905a). On the Electrodynamics of Moving Bodies. First published in *Annalen der Physik, 17*, 891–921; English translation first published in *The Principle of Relativity*, 1923; and reprinted by Dover, New York, pp. 35–66.

Einstein, A. (1905b). Does the Inertia of a Body Depend Upon Its Energy-Content? First published in *Annalen der Physik, 17*, 639–641. English translation first published in *The Principle of Relativity*, 1923; and reprinted by Dover, New York, pp. 67–71.

Einstein, A. (1905c). On a Heuristic Point of View Concerning the Production and Transformation of Light. First published in *Annalen der Physik, xvii*, 132–148. English translation published in Stachel, J. (1998). *Einstein's Miraculous Year*. Princeton University Press, Princeton, New Jersey, pp. 177–198.

Einstein, A. (1911). On the Influence of Gravitation on the Propagation of Light. First published in *Annalen der Physik, 35*, 898–908; English Translation first published in *The Principle of Relativity*, 1923; and reprinted by Dover, New York, pp. 96–108.

Einstein, A. (1916). The Foundation of the General Theory of Relativity. First published in *Annalen der Physik, 49*, 769–822. English Translation first published in *The Principle of Relativity*, 1923; and reprinted by Dover, New York, pp. 109–164.

Einstein, A. (1917). Cosmological Considerations of the General Theory of Relativity. First published in *Sitzungsberichte der Preussischen Akad. D. Wissenschafte*. English translation published in *The Principle of Relativity*, Methuen, 1923; and reprinted by Dover, New York, pp. 175–188.

Einstein, A. (1924). *Relativity — The Special and General Theory*. First derived from Robert W. Lawson's edition published by Methuen & Co. in 1916, then revised in 1924. Published in 2007 by Signature Press Editions, World Publications Group, East Bridgewater, Massachusetts.

Fanchi, J.R. (1977). Quantum Mechanics of Relativistic Spinless Particles. Ph.D. Dissertation, Advisor R.E. Collins. University of Houston, Houston, Texas.

Fanchi, J.R. and R.E. Collins (1978). Quantum Mechanics of Relativistic Spinless Particles. *Foundations of Physics, 8*, 851–877.

Fanchi, J.R. (1986). Parametrizing Relativistic Quantum Theory. *Physical Review A, 34*, 1677–1681.

Fanchi, J.R. (1993a). Review of Invariant Time Formulations of Relativistic Quantum Theories. *Foundations of Physics, 23*, 487–548.

Fanchi, J.R. (1993b). *Parametrized Relativistic Quantum Theory*. Kluwer, Amsterdam.

Fanchi, J.R. (2003). Relativistic Dynamical Theory of Particle Decay and Application to K-Mesons. *Foundations of Physics, 33*, 1189–1205.

Fanchi, J.R. (2011). Manifestly Covariant Quantum Theory with Invariant Evolution Parameter in Relativistic Dynamics. *Foundations of Physics*, *41*, 4–32.

Fanchi, J.R. (2019). Neutrino Flavor Transitions as Mass State Transitions. *Symmetry*, *11*, 948. https://doi.org/10.3390/sym11080948. Open access paper retrieved from https://www.mdpi.com/2073-8994/11/8/948 on March 25, 2022.

Fanchi, J.R. (2021). *Reason, Faith, and Purpose: The Ultimate Gamble*. World Scientific, Singapore.

Fanchi, J.R. (2021). Can particle appearance or disappearance be described by a quantum mechanical theory? *Journal of Physics: Conference Series*, *1956*, 012007. DOI:10.1088/1742-6596/1956/1/012007. Open access paper retrieved from https://iopscience.iop.org/article/10.1088/1742-6596/1956/1/012007 on March 27, 2022.

Fanchi, J.R. (2022). Tachyon Behavior Due to Mass-State Transitions at Scattering Vertices. *Physics*, *4*, 217–228. https://doi.org/10.3390/physics4010016. Open access paper retrieved from https://www.mdpi.com/2624-8174/4/1/16 on March 25, 2022.

Ferrarese, L. (2012). Measuring the Expansion Rate of the Universe. In D. Goodstein (ed.), Chapter 12 in *Adventures in Cosmology*. World Scientific, Singapore.

Feynman, R.P. (1950). Mathematical Formulation of the Quantum Theory of Electromagnetic Interaction. *Physical Review*, *80*, 440–457.

Feynman, R.P., R.B. Leighton, and M. Sands (1963). *The Feynman Lectures on Physics*, 3 volumes. Addison-Wesley, Reading, Massachusetts.

Feynman, R.P. (1998). *Statistical Mechanics: A Set of Lectures*. Addison-Wesley, Reading, Massachusetts.

FitzGerald, G.F. (1889). The Ether and the Earth's Atmosphere. *Science*, *13*(328), 390.

Fock, V.A. (1937). Proper Time in Classical and Quantum Mechanics. Translated by D.H. Delphenich. *Physikalische Zeitschrift der Sowjetunion*, *12*, 404–425.

Fraknoi, A., D. Morrison, and S.C. Wolff (2018). *Astronomy*. OpenStax, Rice University, Houston. Retrieved from https://openstax.org/details/books/astronomy on February 11, 2022.

Frank, A. (2011). *About Time*. Free Press, Simon & Schuster, New York.

Fukugita, M. and C.J. Hogan (2000). Global Cosmological Parameters. *The European Physical Journal C*, *15*, 136–142.

Galilei, Galileo (1632). *Dialogue Concerning the Two Chief World Systems: Ptolemaic and the Copernican*. Translated and with revised notes by Stillman Drake. Series Editor Stephen Jay Gould. Modern Library Paperback Edition, 2001. Random House, New York.

Galilei, Galileo (1638). *Dialogue Concerning Two New Sciences*. Translated by H. Crew and A. De Salvo. Easton Press, Norwalk, Connecticut; printed in 1999.

Gill, T.L., W.W. Zachary, and J. Lindesay (2001). The Classical Electron Problem. *Foundations of Physics*, *31*, 1299–1355.

Glashow, S.L. (1980). Towards a Unified Theory: Threads in a Tapestry. *Reviews of Modern Physics*, *52*, 539–544.

Gleick, J. (2016). *Time Travel — A History*. Penguin Random House, New York.

Goldsmith, D. (1995). *Einstein's Greatest Blunder*. Harvard University Press, Cambridge, Massachusetts.

Greene, B. (2000). *The Elegant Universe*. Vintage Books, New York.

Greene, B. (2020). *Until the End of Time*. Penguin Random House, New York.

Greiner, W., L. Neise, and H. Stöcker (1995). *Thermodynamics and Statistical Mechanics*. Springer-Verlag, New York.

Grøn, Ø. (2020). *Introduction to Einstein's Theory of Relativity*, 2nd Edition. Springer Nature Switzerland AG, New York.

Gross, F. (1993). *Relativistic Quantum Mechanics and Field Theory*. Wiley, New York.

Guth, A. (2002). Inflation and the New Era of High-Precision Cosmology. MIT Physics Annual, MIT, Cambridge, Massachusetts. Retrieved from https://physics.mit.edu/wp-content/uploads/2021/01/physicsatmit_02_cosmology.pdf on February 27, 2022.

Hart-Davis, A. (2011). *The Book of Time*. Octopus Publishing Group, London.

Hartle, J.B. (2021). *Gravity — An Introduction to Einstein's General Relativity*. Cambridge University Press, Cambridge, UK.

Hawking, S. and L. Mlodinow (2010). *The Grand Design*. Bantam Books, New York.

Heisenberg, W. (1927). The actual content of quantum theoretical kinematics and mechanics. English translation of Über den anschaulichen Inhalt der quanten theoretischen Kinematik und Mechanik. *Zeitschrift für Physik*, *43*, 172–198. NASA Technical Memorandum TM-77379 in 1983. Retrieved from https://ia800500.us.archive.org/10/items/nasa_techdoc_19840008978/19840008978.pdf on February 14, 2022.

Helden, A. Van (2021, February 19). Galileo. Encyclopedia Britannica. Retrieved from https://www.britannica.com/biography/Galileo-Galilei on November 13, 2021.

Hesiod (ca. 700 BCE). Works and Days. Written circa 700 BCE. Translated by H.G. Evelyn-White in 1914. Retrieved from https://people.sc.fsu.edu/~dduke/lectures/hesiod1.pdf on November 10, 2021.

History.com Editors (2019, August 23). Hellenistic Greece. Retrieved from https://www.history.com/topics/ancient-history/hellenistic-greece on November 3, 2021.

History.com Editors (2020, November 24). Alexander the Great. Retrieved from https://www.history.com/topics/ancient-history/alexander-the-great on November 3, 2021.

History.com Editors (2021, September 22). Sumer. Retrieved from https://www.history.com/topics/ancient-middle-east/sumer on November 3, 2021.

Horwitz, L.P. (2015). *Relativistic Quantum Mechanics*. Springer, Dordrecht.

Horwitz, L.P. and R.I. Arshansky (2018). *Relativistic Many-Body Theory and Statistical Mechanics*. Morgan & Claypool, San Rafael, California.

Hoyle, F. (1948). A New Model for the Expanding Universe. *MNRAS*, *108*, 372–382.

Huang, K. (2007). *Fundamental Forces of Nature*. World Scientific, Singapore.

Hubble, E. (1929). A Relation Between Distance and Radial Velocity Among Extra-Galactic Nebula. *Proceedings of the National Academy of Sciences*, *15*, 168–173.

Hubble, E. and M.L. Humason (1931). The Velocity-Distance Relation among Extra-Galactic Nebulae. *Astrophysical Journal*, *74*, 43–80.

Hume, D. (1748). *Enquiry Concerning Human Understanding*. Online version by J. Bennett, 2017. Retrieved from https://www.earlymoderntexts.com/assets/pdfs/hume1748.pdf on November 21, 2021.

Huygens, C. (1656). On the Motion of Bodies Resulting from Impact. Translation by M.S. Mahoney, 1995. Retrieved from https://www.princeton.edu/~hos/Mahoney/texts/huygens/impact/huyimpct.html on November 22, 2021.

Huygens Patent (2017, June). June 16, 1657: Christiaan Huygens Patents the First Pendulum Clock. This Month in Physics History. APS News. Retrieved from https://www.aps.org/publications/apsnews/201706/history.cfm on November 14, 2021.

Jackson, J.D. (1987, May). Thc Impact of Special Relativity on Theoretical Physics. *Physics Today*, *40*, 34–42.

Jackson, J.D. (1999). *Classical Electrodynamics*, 3rd Edition. Wiley, New York.

Jones, A.R. (2020, May 19). Ptolemaic system. Encyclopedia Britannica. Retrieved from https://www.britannica.com/science/Ptolemaic-system on November 6, 2021.

Joyce, J. (1939). *Finnegans Wake*. Faber & Faber Limited, London. Retrieved from https://archive.org/details/in.ernet.dli.2015.463592/mode/2up on February 24, 2022.

Kaku, M. (1993). *Quantum Field Theory*. Oxford University Press, Oxford, UK.

Kaku, M. (2021). *The God Equation*. Doubleday, New York.

Keesing, F.M., M. Gimbutas, H.L. Movius, R.J. Braidwood, R.M. Adams, and R. Pittioni (2020, November 10). Stone Age. Encyclopedia Britannica. Retrieved from https://www.britannica.com/event/Stone-Age on October 31, 2021.

Kisslinger, L.S. (2017). *Astrophysics and the Evolution of the Universe*, 2nd Edition. World Scientific, Singapore.

Kolb, E.W. and M.S. Turner (2000). The Pocket Cosmology. *The European Physical Journal C*, *15*, 125–132.

Kragh, H. (1996). *Cosmology and Controversy: The Historical Development of Two Theories of the Universe*. Princeton University Press, Princeton, New Jersey.

Kuhn, T.S. (1970). *The Structure of Scientific Revolutions*, 2nd Edition. University of Chicago Press, Chicago.

Kumar, M. (2008). *Quantum*. Icon Books, Thriplow, Cambridge.

Land, M. and L.P. Horwitz (2020). *Relativistic Classical Mechanics and Electrodynamics*. Morgan & Claypool, Williston, Vermont.

Lang, K.R. (2018). *A Brief History of Astronomy and Astrophysics*. World Scientific, Singapore.

Lanza, R. and M. Pavšič, with B. Berman (2020). *The Grand Biocentric Design: How Life Creates Reality*. BenBella Books, Dallas, Texas.

Lautrop, B. (2005). *Physics of Continuous Matter*. Institute of Physics Publishing, Bristol, UK.

Leavitt, H.S. (1908). 1777 Variables in the Magellanic Clouds. *Annals of Harvard College Observatory*, *60*(4), 87–108.3.

Leavitt, H.S. and E.C. Pickering (1912). Periods of 25 Variable Stars in the Small Magellanic Cloud. *Harvard College Observatory Circular*, *173*, 1–3, accessed December 24, 2020 from https://silentskyplay.tumblr.com/post/71219744368/henrietta-swan-leavitts-actual-1912-paper-wherein.

Leinaas, J.M. (2019). *Classical Mechanics and Electrodynamics*. World Scientific, Singapore.

Lemaître, G. (1933). La formation des Nebuleuses dans l'Univers en Expansion. *Comptes Rendus*, *196*, 903–904. As translated in Helge Kragh, *Cosmology and Controversy: The Historical Development of Two Theories of the Universe* (1996), Princeton University Press, Princeton, New Jersey, pp. 22–79.

Liddle, A. (1999). *An Introduction to Modern Cosmology*. Wiley, New York.

Linde, A. (1987, September). Particle Physics and Inflationary Cosmology. *Physics Today*, *40*, 61–68.

Lindley, D. (1987, December). Cosmology from Nothing. *Nature*, *330*, 603–604.

Ludvigsen, M. (1999). *General Relativity*. Cambridge University Press, Cambridge, UK.

Mach, E. (1919). *Science of Mechanics*, 4th Edition. Translated from the German by T.J. McCormack. The Open Court Publishing Company, London. Retrieved from https://ia804508.us.archive.org/7/items/scienceofmechan-i005860mbp/scienceofmechani005860mbp.pdf on January 15, 2022.

Malkan, M.A. and B. Zuckerman, (eds.) (2020). *Origin and Evolution of the Universe*, 2nd Edition. World Scientific, Singapore.

Maoz, D. (2007). *Astrophysics in a Nutshell*. Princeton University Press, Princeton, New Jersey.

Mendeleev, D. (1869). On the Relationship of the Properties of the Elements to Their Atomic Weights. Translation from the German by Carmen Giunta. *Zeitschrift für Chemie, 12*, 405–406.

Mendeleev, D. (1889). The Periodic Law of the Chemical Elements. *Journal of the Chemical Society, Translations*, *55*, 634–656. DOI:10.1039/CT889 5500634.

Mendeleev, D. (1905). Grouping of the Elements and the Periodic Law. From *The Principles of Chemistry* by D. Mendeleev. Translated by George Kemansky, edited by T.H. Pope, Volume II. Longmans, Green and Co., New York and Bombay, pp. 10–34. Reproduced in Wolff, P. (1967). *Breakthroughs in Chemistry*. New American Library, New York, pp. 231–248.

Minkowski, H. (1908). Space and Time. A translation of an address delivered at the 80th Assembly of German Natural Scientists and Physicians, Cologne, 21 September 1908; published in *The Principle of Relativity*, 1923; and reprinted by Dover, New York, pp. 73–91.

Misner, C.W., K.S. Thorne, and J.A. Wheeler (1973). *Gravitation*. W.H. Freeman, San Francisco.

Mitchell, S.W. (2003, August). The Classic: The History of Instrumental Precision in Medicine. *Clinical Orthopaedics and Related Research*, *413*, 11–18. Retrieved from https://journals.lww.com/clinorthop/Fulltext/2003/08000/The_Classic__The_History_of_Instrumental_Precision.3.aspx on November 16, 2021.

Morin, D. (2007). *Introduction to Classical Mechanics*. Cambridge University Press, Cambridge, UK.

Moring, G.F. (2002). *The Complete Idiot's Guide to Theories of the Universe*. Penguin Group, New York City.

Muller, R.A. (2016). *Now: The Physics of Time*. W.W. Norton and Co., New York.

NASA-COBE (2009). COBE Satellite marks 20th Anniversary, published online November 17, 2009, by the National Aeronautics and Space Administration, accessed October 13, 2019 from https://www.nasa.gov/topics/universe/features/cobe_20th.html.

NASA-WMAP (2022). Timeline of the Universe. National Aeronautics and Space Administration (NASA) WMAP Science Team. Retrieved from https://wmap.gsfc.nasa.gov/media/060915/index.html on February 28, 2022.

NASA-WMAP Science Team (2022). Cosmology: The Study of the Universe. National Aeronautics and Space Administration (NASA), Wilkinson Microwave Anisotropy Probe. Retrieved from https://wmap.gsfc.nasa.gov/universe/ on February 28, 2022.

Newton, I. (1687). *Mathematical Principles of Natural Philosophy*. Published in London; also known as *Principia*. Easton Press, Norwalk Connecticut; printed in 2000.

Newton, I. (1692–1693). *Four Letters from Sir Isaac Newton to Doctor Bentley containing Some Arguments in Proof of a Deity*. Published in London;

printed for R. and J. Dodsley, Pall Mall, 1756. Retrieved from https://ia801901.us.archive.org/2/items/FourLettersFromSirSaacNewtonToDoctorBentleyContainingSomeArgumentsInProofOfADEITY/Four%20Letters%20from%20Sir%20saac%20Newton%20to%20Doctor%20Bentley%20containing%20some%20Arguments%20in%20Proof%20of%20a%20DEITY.pdf on January 10, 2022.

Newton, R.G. (2004). *Galileo's Pendulum*. Harvard University Press, Cambridge, Massachusetts.

Nussbaumer, H. (2014). Einstein's Conversion from His Static to an Expanding Universe. *The European Physical Journal H, 39*, 37–62.

Odenwald, S. (2019). *Cosmology*. Arcturus Holdings, London.

Padmanabhan, T. (2006). *An Invitation to Astrophysics*. World Scientific, Singapore.

Pais, A. (1982). *Subtle Is the Lord...*. Oxford University Press, Oxford, UK.

Palmer, D. (2009). *Evolution — The Story of Life*, illustrated by P. Barrett. University of California Press, Berkeley, California.

Pauli, W. (1958). *Theory of Relativity*. Dover, New York.

Pavšič, M. (2001). *The Landscape of Theoretical Physics: A Global View*. Kluwer Academic Publishers, Dordrecht.

Pavšič, M. (2020). *Stumbling Blocks Against Unification*. World Scientific, Singapore.

Peebles, P.J.E. (1993). *Principles of Physical Cosmology*. Princeton University Press, Princeton, New Jersey.

Perkowitz, S. (2021, June 1). Relativity. Encyclopedia Britannica. Retrieved from https://www.britannica.com/science/relativity on January 28, 2022.

Perlov, D. and A. Vilenkin (2017). *Cosmology for the Curious*. Springer Nature Switzerland AG, Cham, Switzerland.

Pickrell, J. (2006, September 4). Timeline: Human Evolution. New Scientist. Retrieved from https://www.newscientist.com/article/dn9989-timeline-human-evolution/ on October 31, 2021.

Planck Collaboration (2020). Planck 2018 Results. *Astronomy and Astrophysics, 641*, 1–67. Retrieved from https://doi.org/10.1051/0004-6361/201833910.

Plato (1991). The Allegory of the Cave. Book VII of *The Republic*, Socratic dialogue authored by Plato and published circa 375 BCE. Translated by B. Jowett. Vintage, pp. 253–261. Retrieved from https://www.studiobinder.com/blog/platos-allegory-of-the-cave/ on February 4, 2021.

Reif, F. (1965). *Fundamentals of Statistical and Thermal Physics*. McGraw-Hill, New York.

Riess, A.G., *et al.* (2016). A 2.4% Determination of the Local Value of the Hubble Constant. *The Astrophysical Journal*, *826*, 1–31. DOI:10.3847/0004-637X/826/1/56.

Rooney, A. (2015). *The Story of Physics*. Arcturus Publishing, London.

Rovelli, C. (2008). Quantum gravity. *Scholarpedia*, *3*(5), 7117.

Rovelli, C. (2015). Aristotle's Physics: A Physicist's Look. *Journal of the American Philosophical Association, 1*, 23–40. DOI:10.1017/apa.2014.11.

Rovelli, C. (2017). *The Order of Time*. Riverhead Books — Penguin Random House, New York.

Sambursky, S. (1975). *Physical Thought from the Presocratics to the Quantum Physicists*. Pica Press, New York.

Schmitz, W. (2019). *Particles, Fields and Forces*. Springer Nature Switzerland AG, Cham, Switzerland.

Singh, S. (2004). *Big Bang*. Fourth Estate, HarperCollins, London.

Slipher, V. (1915). Spectrographic Observations of Nebulae. Report of the 17th Meeting of the American Astronomical Society, pp. 21–24. Retrieved from https://www.roe.ac.uk/~jap/slipher/slipher_1915.pdf on February 5, 2022.

Smithsonian Institution (2020). Human Family Tree. Last updated December 9, 2020. Retrieved from https://humanorigins.si.edu/evidence/human-family-tree on November 1, 2021.

Smolin, L. (2013). *Time Reborn*. Houghton Mifflin Harcourt, New York.

Sommerfeld, A. (1923). *The Principle of Relativity*. English Translation, first published in Methuen and Company, 1923; and reprinted by Dover, New York.

SPACE Editors (1992, November 7). Vatican Admits Galileo was right. New Scientist. Retrieved from https://www.newscientist.com/article/mg13618460-600-vatican-admits-galileo-was-right/ on November 10, 2021.

Stachel, J. (1998). *Einstein's Miraculous Year*. Princeton University Press, Princeton, New Jersey.

Stroll, A. and R.H. Popkins (1972). *Introductory Readings in Philosophy*. Holt, Rinehart and Winston, New York.

Stueckelberg, E.C.G. (1941a). La signification du temps propre en mecanique endulatoire. *Helvetica Physica Acta, 14*, 322–323.

Stueckelberg, E.C.G. (1941b). Remarque a propos de la creation de paires de particules en theorie de relativite. *Helvetica Physica Acta, 14*, 588–594.

Stueckelberg, E.C.G. (1942). La mecanique du point materiel en theorie de relativite et en theorie des quanta. *Helvetica Physica Acta, 15*, 23–37.

Teerikorpi, P., M. Valtonen, K. Lehto, H. Lehto, G. Byrd, and A. Chernin (2019). *The Evolving Universe and the Origin of Life*, 2nd Edition. Springer Nature Switzerland, Gewerbestrasse, Switzerland.

Tudge, C. (2000). *The Variety of Life*. Oxford University Press, Oxford, UK.

Tyson, N.D. and D. Goldsmith (2004). *Origins*. W.W. Norton, New York.

Uffink, J. (2004). Boltzmann's Work in Statistical Physics. The Stanford Encyclopedia of Philosophy (Winter 2004 Edition), In E.N. Zalta (ed.). Retrieved from https://plato.stanford.edu/archives/spr2017/entries/statphys-Boltzmann/ on March 5, 2022.

Urone, P.P., R. Hinrichs, K. Dirks, and M. Sharma (2020). *College Physics*. OpenStax, Rice University, Houston. Retrieved from https://openstax.org/details/books/college-physics on March 5, 2022.

Vaidman, L. (2021). Many-Worlds Interpretation of Quantum Mechanics. The Stanford Encyclopedia of Philosophy (Fall 2021 Edition), In E.N. Zalta (ed.). Retrieved from https://plato.stanford.edu/archives/fall2021/entries/qm-many worlds/ on February 20, 2022.

Veltman, M. (2018). *Facts and Mysteries in Elementary Particle Physics*, Revised Edition. World Scientific, Singapore.

Ventrudo, B. (2012). The Expanding Universe. One-Minute Astronomer. Part 1 retrieved from https://oneminuteastronomer.com/5447/vesto-slipher/ on February 5, 2022; Part 2 retrieved from https://oneminuteastronomer.com/5478/edwin-hubble/ on February 5, 2022; and Part 3 retrieved from https://oneminuteastronomer.com/5576/expanding-universe/ on February 5, 2022.

Violatti, C. (2014, July 18). Stone Age. World History Encyclopedia. Retrieved from https://www.worldhistory.org/Stone_Age/ on October 31, 2021.

Weinberg, S. (1993). *The First Three Minutes*, Updated Edition. Basic Books, New York.

Weinberg, S. (2008). *Cosmology*. Oxford University Press, Oxford, UK.

Weinberg, S. (2015). *Lectures on Quantum Mechanics*, 2nd Edition. Cambridge University Press, New York.

Weinberg, S. (2021). *Foundations of Modern Physics*. Cambridge University Press, New York.

Weinstein, S. and D. Rickles (2019). Quantum Gravity. The Stanford Encyclopedia of Philosophy (Fall 2021 Edition), In E.N. Zalta (ed.). Retrieved from https://plato.stanford.edu/archives/fall2021/entries/quantum-gravity/ on March 14, 2022.

Wells, H.G. (1895). *The Time Machine*. Easton Press, Norwalk, Connecticut, printed in 2002.

Wells, H.G. (1898). *The War of the Worlds*. Easton Press, Norwalk, Connecticut, printed in 1964.

Wheeler, J.A. (1990). *A Journey into Gravity and Spacetime*. Scientific American Library, W.H. Freeman, New York.

Wheeler, J.A. (2010). *Geons, Black Holes, and Quantum Foam: A Life in Physics*. W.W. Norton & Company, New York.

Wolff, P. (1965). *Breakthroughs in Physics*. New American Library, New York.

Wolff, P. (1967). *Breakthroughs in Chemistry*. New American Library, New York.

Zwicky, F. (1933). Die Rotverschiebung von extragalaktischen Nebeln. *Helvetica Physica Acta*, *6*, 110–127, in German.

Zwicky, F. (1937). On the Masses of Nebulae and of Clusters of Nebulae. *Astrophysical Journal*, *86*, 217–246. DOI:10.1086/143864.

Zyla, P.A., *et al.* (2020). Particle Data Group, Review of Particle Physics. *Progress of Theoretical and Experimental Physics*, *2020*(8), 083C01, accessed December 7, 2020 from https://doi.org/10.1093/ptep/ptaa104, see also https://academic.oup.com/ptep/article/2020/8/083C01/5891211.

Index

J

K

L

M

CPSIA information can be obtained
at www.ICGtesting.com
Printed in the USA
JSHW012231270323
39469JS00004B/25

9 781800 613348